Study Guide For Radiation Oncology Physics Board Exams

Brian D. Berman, M.S., D.A.B.R.

Edited by Bruce R. Thomadsen, Ph.D., D.A.B.R., D.A.B.M.P., D.A.B.H.P.

Medical Physics Publishing
Madison, WI

Copyright © 2000 by Brian D. Berman

All rights reserved. No part of the publication may be reproduced or distributed in any form or by any means without the prior written permission of the publisher.

Second printing February 2002

Library of Congress Catalog Card Number: 00-100718
ISBN 0-944838-94-4

Published by:
Medical Physics Publishing
4513 Vernon Boulevard
Madison, WI 53705-4964
(608) 262-4021
1-800-442-5778
Email: mpp@medicalphysics.org
Web Site: www.medicalphysics.org

Information in this study guide is provided for instructional use only. Because of the possibility for human error and the potential for change in the medical sciences, the reader is strongly cautioned to verify all information with an independent source. The authors, publisher, and printer cannot assume any responsibility for validity of all materials or for any damage or harm incurred as a result of the use of this information.

Printed in the United States of America

Contents

Preface xiii

CHAPTER 1: BASIC PHYSICS — 1

- SI Prefixes 1
- SI Quantities 1
- Constants 2
- Physical Relations 2
  - Mathematical 2
  - Motion 3
  - Geometric 3
  - Energy 3
- Statistics and Counting 3
  - Means 3
  - Probability Distributions 4
  - Radioactive Decay Counting 4
- Temperature and Pressure 5
  - Constants 5
  - Relations 5
- The Atom 5
  - Constants 5
  - Representation: $^{A}_{Z}X$ 5
  - Descriptors 6
  - Binding Energy (BE) or "Mass Defect" 6
- Electromagnetic Radiation 6

CHAPTER 2: BASIC NUCLEAR AND ATOMIC PHYSICS — 7

- Radioactivity 7
  - Natural 7
  - Artificial 7

Exponential Behavior ... 7
Activity ... 8
General ... 8
Units ... 8
Half-Life And Average Life ... 8
Half-Life (t_H) ... 8
Average or Mean Life (t_A) ... 8
Effective Half-Life (t_E) ... 9
Radioactive Equilibrium ... 9
Radioactive Daughter ... 9
Metastable State ... 10
Transient Equilibrium ... 10
Secular Equilibrium ... 10
Modes of Radioactive Decay ... 11
Isomeric Decay ... 11
Alpha Decay ... 12
Isobaric Decay ... 12
Isomeric Transition ... 13
Atomic Physics ... 13
Auger Emission ... 13
Characteristic Radiation ... 14

CHAPTER 3: PRODUCTION OF X RAYS ... 15
X-Ray Production ... 15
Bremsstrahlung Radiation ... 15
Characteristic X-Rays ... 15
X-Ray Unit ... 16
Basic Circuit Principle ... 16
Anode ... 16
Cathode ... 17
Heat Production of Diagnostic Tubes ... 17
Heat Units (HUs) ... 17
Energy Deposited ... 17

Focal Spot	18
X-Ray Tube Operating Characteristics	18
Filters	18
Inherent Filtration	19
Diagnostic	19
Therapy – Graded (Thoraeus)	19
Energy Spectra	19
"Thin Target" Approximation	19
"Thick Target" Approximation	20
X-Ray Image	20
Image Quality	20
Modulation Transfer Function (MTF)	20
Computed Tomography (CT)	21
CT Number	21
Hounsfield Unit (HU)	21
Mammography	21

CHAPTER 4: THERAPY RADIATION GENERATORS — 23

Kilovoltage Units	23
Grenz	23
Contact Therapy	23
Superficial	23
Orthovoltage	23
Megavoltage Units	24
Van de Graaff	24
Betatron	24
Cyclotron	24
Microtron	24
Linear Accelerator (LINAC)	24
Cobalt-60 Units	26
Source	26
Dosimetric Characteristics	27

CHAPTER 5: INTERACTIONS OF ELECTRONS AND CHARGED PARTICLES WITH MATTER 29

Ionization and Excitation 29
Ionization 29
Excitation 29

Collisions 29
Elastic 30
Inelastic 30

Energy Losses 30
Stopping Power (S) 31
Restricted Stopping Power (L) 31
Range 32
Bragg Peak 32
Bremsstrahlung Yield (B) 33

Polarization ("Density") Effect 33

Electron Energy Spectrum 34
Most Probable Incident Energy 34
Mean Incident Energy 34
Energy at Depth 34

CHAPTER 6: INTERACTION OF IONIZING RADIATION WITH MATTER 35

Interaction of Photon Beam 35
Description 35
Exponential Attenuation 35
Attenuation Coefficients 36
Energy Coefficients 36
Half-Value Layer (HVL or X_H) 37
Spectrum 37

Coherent Scattering 37
Thomsom (Classical) 38
Rayleigh (Coherent) 38

Photoelectric Interaction 38

Incoherent or Compton Scattering	39
Energy-Angle Relationships	40
Klein-Nishina Coefficients	41
Pair Production	42
Photodisintegration	43
Relative Importance of Interactions	43

CHAPTER 7: MEASUREMENT OF IONIZING RADIATION AND ABSORBED DOSE — 45

The Roentgen	45
Exposure (X) = dQ/dm	45
f-factor (f)	45
Chambers	46
Chamber Types	46
Chamber Wall	47
Stem Effect	48
Desirable Chamber Characteristics	48
Ion Collection	48
Saturation	48
Recombination	48
Temperature and Pressure	49
W/e Factor	49
Electronic Equilibrium (EE)	49
Polarity Effects	50
Energy Transfer	51
Kerma (Kinetic Energy Released in the Medium)	51
Absorbed Dose	51
Energy Transfer and Dose Curves	51
Bragg-Gray Cavity Theory	52
Bragg-Gray Relationship	52
Spencer-Attix Formulation of Bragg-Gray Relationship	53

Absorbed Dose in a Medium (TG-21 Calibration Protocol)	53
Absorbed Dose from Photons	53
Absorbed Dose from Electrons	55
Absorbed Dose in a Medium (TG-51 Calibration Protocol)	56
Absorbed Dose from Photons	56
Absorbed Dose from Electrons	57
Other Means To Measure Absorbed Dose	59
Absolute Measurement of Dose	59
Relative Measurement of Dose	59
Typical Dose Calibration Uncertainties	60

CHAPTER 8: RADIATION THERAPY TREATMENT PLANNING   61

Dose Distribution	61
Depth of Maximum Dose (d_{max})	61
Photon Equivalent Squares	61
Geometric Penumbra	62
Flatness and Symmetry	62
Electron Dose Distribution	62
Photon Beam Correction Factors	63
Scatter Factors	63
Transmission Factors	63
Dosimetric Calculations	64
Fractional Depth Dose (FDD)	64
Tissue-Air Ratio (TAR)	65
Peakscatter Factor (PSF)	66
Scatter-Air Ratio (SAR)	66
Tissue-Maximum Ratio (TMR)	67
Linear Accelerator Monitor Unit Calculations	68
SSD Calculations	68
SAD Calculations	69

Photon Dose Calculation Algorithms 69

    Dose Computation Based on Correcting Measured Dose 69

    Model-Based Dose Calculation 70

Pencil-Beam Electron Dose Calculation Algorithm 70

Field Matching for Photon Beams 71

    Separation of Adjacent Fields 71

    Craniospinal Fields 72

    Matching of Fields without Divergence 72

    Wedged Fields 72

Heterogeneity Corrections 73

    One-dimensional Heterogeneity Corrections for Photon Beams 73

    Three-dimensional Heterogeneity Corrections for Photon Beams 74

    Heterogeneity Corrections for Electron Beams 75

CHAPTER 9: BRACHYTHERAPY PHYSICS   77

Applications 77

    Interstitial Implants 77

    Intracavitary Insertions 77

    Surface Applications 77

Treatment Types 78

    Temporary 78

    Permanent 78

Source Strength Specification 78

    Exposure Rate 78

    Apparent Activity 79

    Radium Equivalent 79

    Air Kerma Strength 79

Radionuclides 80

    Radium-226 80

    Cesium-137 80

    Gold-198 80

    Iridium-192 81

Iodine-125	81
Palladium-103	81
Strontium-90	82

Localization ... 82
- Orthogonal Films ... 82
- "Stereo" Films ... 82

Dosimetry ... 83
- Conventional Point Source Approximation ... 83
- Sievert Integral ... 83
- Task Group 43: Dosimetry Formalism for Interstitial Sources (^{192}Ir, ^{125}I, ^{103}Pd) ... 84

Implant Dosing Systems ... 85
- Interstitial Systems ... 85
- Intercavitary Systems (Cervical Cancer) ... 86

CHAPTER 10: RADIATION PROTECTION — 87

Definitions ... 87
- Equivalent Dose ... 87
- Effective Dose ... 87
- Committed Equivalent Dose ... 88
- Committed Effective Dose ... 88

Linear Energy Transfer (LET) ... 88

Effective Dose Limits ... 89
- Areas ... 89
- Occupational Limits (NCRP Report No. 116) ... 89
- Public Limits (NCRP Report No. 116) ... 89

Shielding ... 89
- Photon Shielding (NCRP Reports No. 49 and No. 51) ... 89
- Neutron Shielding ... 91

Monitoring Instruments ... 92
- Film ... 92
- Solid-State Detectors ... 92
- Gas-Filled Chambers ... 92

CHAPTER 11: RADIOBIOLOGY 95

Biological Effects of Radiation 95
- Biological Effects 95
- Radiobiological Equivalent (RBE) 95
- Oxygen Enhancement Ratio (OER) 96

Cell Survival 96
- Cell Radiosensitivity 96
- Alpha-Beta Model of Cell Survival: $S = e^{-(\alpha \cdot D + \beta \cdot D^2)}$ 96
- Cell Survival Curve 97

Terms And Acronyms 99

Suggested Readings 101

Preface

This study guide is intended to assist the board exam candidate in the field of radiation oncology physics in preparing for his or her written exams. It covers both general physics and therapeutic radiological physics. The material found within this text is intended to be a "guide" and an aide to those studying for the physics board exams, not a replacement for the rigorous studying required to pass such exams. The user should regard it as an incomplete outline of fundamentals by which you can recognize weaknesses in your understanding of the field and determine specific areas that require further study.

This study guide does not contain problems, nor does it include any explicit answers to questions found on the board exams. It contains basic principles in the medical physics field, the knowledge of which will deepen your rudimentary understanding of the subject, benefit your on-the-job performance, and improve your board exam scores.

Every effort has been made to ensure that the information herein is accurate and true; however, errors may still be present. All information contained in the guide should be verified and supplemented with information found in other relevant texts.

I am deeply indebted to Bruce Thomadsen for his assistance in completing this project, and his help in creating a comprehensive and accurate study guide. Without his detailed proofreading and prudent editing, this work would certainly be less complete and in many respects flawed. I would also like to acknowledge Tom Yip for his time and effort in creating the figures used in this guide.

CHAPTER 1

BASIC PHYSICS

SI Prefixes

Name	Notation	Name	Notation
deci (d)	10^{-1}	deka (da)	10^{+1}
centi (c)	10^{-2}	hecto (h)	10^{+2}
milli (m)	10^{-3}	kilo (k)	10^{+3}
micro (µ)	10^{-6}	mega (M)	10^{+6}
nano (n)	10^{-9}	giga (G)	10^{+9}
pico (p)	10^{-12}	tera (T)	10^{+12}
femto (f)	10^{-15}	peta (P)	10^{+15}

SI Quantities

Basic Quantity	SI Unit	Useful Relation
length (l)	meter (m)	2.54 cm / in
mass (m)	kilogram (kg)	453.6 g / lb
time (t)	second (s)	86,400 s / day
current (I)	ampere (A)	Amp = C/s
temperature (T)	kelvin (K)	22 °Celsius = 295 K
luminance (L)	candela (cd)	—

Derived Quantity	SI Unit	Derivation	Useful Relation
frequency (v)	hertz (Hz)	$1 \text{ Hz} = 1 \text{ s}^{-1}$	period = $1/v$
force (F)	newton (N)	$F = m \cdot a$ (kg·s^2)	1 N = 0.225 lb
energy (E)	joule (J)	$E = F \cdot l$ (N·m)	4.19 J = 1 cal
pressure (P)	pascal (Pa)	Pressure = F / l^2 (N/m^2)	1.01×10^5 Pa = 1 atm
power (P)	watt (W)	Power = E / t (J/s)	746 W = 1 hp
charge (Q)	coulomb (C)	$Q = I \cdot t$ (A·s)	$Q(e-) = 1.6 \times 10^{-19}$ C
potential (V)	volt (V)	$V = E / Q$ (W/A)	1 eV = 1.60×10^{-19} J
resistance (R)	ohm (Ω)	$R = V / I$ (V/A)	$P = V^2 / R = I^2 \cdot R$
capacitance (C)	farad (F)	$C = Q / V$ (A·s/V)	$V = V_0 \cdot e^{-t/RC}$

Constants

Speed of light (c) = 3.00×10^8 m/s

Acceleration of gravity (g) = 9.8 m/s^2

Planck constant (h) = 6.63×10^{-34} J·s

Boltzmann constant (k) = 1.38×10^{-23} J/K

Gravitational constant (G) = 6.67×10^{-11} m^3/s^2·kg

Physical Relations

Mathematical

1. Quadratic formula: $x = \left[\dfrac{-b \pm \sqrt{(b^2 - 4 \cdot a \cdot c)}}{2 \cdot a} \right]$

    when

    $a \cdot x^2 + b \cdot x + c = 0$.

2. Pythagorean theorem: $a^2 + b^2 = c^2$

    where

    c is the hypotenuse of a right triangle and

    a and b are the arms of the triangle.

Motion

1. Angular momentum = $m \cdot v \cdot r$

   where

   m is mass in kg,

   v is velocity in m/s, and

   r is the radius of motion in m.

2. Centripetal force = $m \cdot v^2 / r$

3. Gravitational force = $G \cdot m_1 \cdot m_2 / r^2$

Geometric

1. Volume of sphere = $4/3 \cdot \pi \cdot r^3$

2. Volume of ellipsoid = $1/8 \cdot \pi \cdot a \cdot b \cdot c$

3. Surface area of sphere = $4 \cdot \pi \cdot r^2$

4. Solid angle (Ω) = Area $/ 4 \cdot \pi \cdot r^2$ (or Area/r^2 in steradians)

Energy

1. Kinetic Energy (KE) = $1/2 \cdot m \cdot v^2$

2. Relativistic Energy (E) = $m \cdot c^2 = m_0 \cdot c^2 + KE = \dfrac{E_0}{\sqrt{1 - \dfrac{v^2}{c^2}}}$

3. Relativistic Mass (m) = $\dfrac{m_0}{\sqrt{1 - \dfrac{v^2}{c^2}}}$

Statistics and Counting

Means

1. Arithmetic mean = $\dfrac{(X_1 + X_2)}{2}$

2. Harmonic = $\dfrac{2 \cdot X_1 \cdot X_2}{(X_1 + X_2)}$

3. Geometric = $\sqrt{(X_1 \cdot X_2)}$ (used for radioactive decay and exponential attenuation)

Probability Distributions

1. Binomial probability: $P_B(x) = \left[\dfrac{N!}{X! \cdot (N-X)!}\right] \cdot P^x \cdot Q^{N-x}$

   where

   P_B is the probability of X successes in N trials,

   P is the probability of success in one trial,

   Q is the probability of failure in one trial, and

   the standard deviation is $(\rho) = \sqrt{(N \cdot P \cdot Q)}$

2. Poisson probability: $P_P(x) = \dfrac{A^x \cdot e^{-A}}{x!}$

   where

   A is the expected value (or average),

   P_P is the probability of observing the value X,

   standard deviation is $(\sigma) = \sqrt{A}$, and

   probable error is $(p) = 0.67 \cdot \rho$.

3. Normal (Gauss) error integral: $P_G(t\sigma) = erf(t) = \left[\dfrac{1}{\sqrt{(2\pi)}}\right] \cdot \left[\int_{-t}^{t} e^{-x^2/2} dx\right]$

   where

   P_G is the probability of a number lying between $-t \cdot \sigma$ and $+t \cdot \sigma$ of a normal distribution.

Radioactive Decay Counting

1. Insensitive period of a radiation counter after a measurement pulse has been received or "dead time": $(\tau) = \dfrac{(N_A + N_B - N_{AB})}{(2 \cdot N_A \cdot N_B)}$

   where

   N_A and N_B are the observed counts per second of two sources A and B and

   N_{AB} is the observed counts per second of sources A and B together.

2. Paralyzable counter: $N_0 = T \cdot e^{-T\tau}$

   where

   N_0 is the observed counts and

   T is the true counts.

3. Non-paralyzable counter: $N_0 = T / (1 - T \cdot \tau)$.

Temperature and Pressure

Constants

1. Triple point of water (T_3) = 273.15 K at 101.3 kPa
2. Universal gas constant (R) = 8.31 J/mol · K

Relations

1. Energy (E) = $3/2 \cdot k \cdot T$
2. $P \cdot V$ = $n \cdot R \cdot T$
3. P = $P_0 \, e^{-(g \cdot h/Pa)}$
4. Standard temperature and pressure (STP):
    T = 273.15, P = 760 mmHg = 101.3 kPa

The Atom

Constants

1. Avogadro's constant (N_A) = 6.02×10^{23} mol^{-1}
2. Bohr radius (r_B) = 5.29×10^{-11} m
3. Nuclear radius (r_n) = 10^{-14} m
4. Atomic Mass Unit (amu) = 1/12 mass of ^{12}C = 1.66×10^{-27} kg = 931 MeV
5. Electron rest mass (m_e) = 9.11×10^{-31} kg = 0.511 MeV
6. Proton rest mass (m_p) = 1.67×10^{-27} kg = 936 MeV
7. Neutron rest mass (m_n) = 1.68×10^{-27} kg = 942 MeV
8. Fine structure constant (α) ≈ 1/137

Representation: $^A_Z X$

1. X = Element symbol
2. Z = Atomic number = Number of protons
3. A = Mass number = Number of protons and neutrons

Descriptors

1. Neutral atom: When number of electrons = Z.
2. Excited atom: When number of electrons ≠ Z.
3. Isotopes: Atoms with the same Z but different A.
4. Isomers: Atoms with the same Z and A but different energy state.
5. Isobars: Atoms with the same A but different Z.
6. Isotones: Atoms with the same number of neutrons (A-Z) but different Z.

Binding Energy (BE) or "Mass Defect"

BE is the energy required to separate nucleons from the nucleus.

1. BE = (mass of the nucleus) − (mass sum of constituents).
2. BE is a measure of the stability of a nucleus.

Electromagnetic Radiation

Quantum nature of radiation: $E = h \cdot \nu$

Wave nature of radiation: $c = \lambda \cdot \nu$

CHAPTER 2

BASIC NUCLEAR AND ATOMIC PHYSICS

Radioactivity

Natural

A nucleus that is in an excess energy state can emit energy in the form of particles or electromagnetic radiation in order to return to a more stable state.

Artificial

High-energy particle bombardment of atoms can create new radioactive isotopes when the nuclei absorb these projectiles or when they cause the nuclei to eject a particle.

Exponential Behavior

$$N(t) \propto dN/dt \quad \rightarrow \quad N(t) = \lambda \cdot dN/dt \quad \rightarrow \quad N(t) = N_0 \, e^{-\lambda \cdot t}$$

where

$N(t)$ is a function whose rate of change over a time t is proportional to the function itself and

λ is a constant of proportionality called the decay constant.

Activity

General

1. Activity (A): The rate of decay or change in number of atoms of a radionuclide over time.
2. $A(t) = -\lambda \cdot N(t) = N_0 \cdot \lambda \cdot e^{-\lambda \cdot t} = A_0 \cdot e^{-\lambda \cdot t}$
3. The constant of proportionality (λ), or decay constant, is representative of the fractional change in the number of unstable atoms that occurs in unit time.

Units

1. 1 Becquerel (Bq) = 1 disintegration per second (dps)
2. 1 Curie (Ci) = 3.7×10^{10} dps ≈ activity from 1 gram radium

Half-Life and Average Life

Half-Life (t_H)

The time required to reduce the number of radioactive atoms to half their initial number through radioactive decay.

1. $A = 1/2 \cdot A_o$ when $t = t_H$
2. $t_H = \ln 2 / \lambda = 0.693 / \lambda$

Average or Mean Life (t_A)

The time required for a hypothetical source with a constant activity of A_0 to produce the same number of disintegrations as an exponentially decaying source with initial activity of A_0 that completely decays.

1. $t_A = 1/\lambda = 1.44 \cdot t_H$
2. t_A is the time to decay averaged over all nuclei in a sample.

Effective Half-Life (t_E)

The half-life of an ingested radioisotope that undergoes a physical decay with half-life (t_H) as well as a biological decay half-life (t_B), which is the rate at which the radioisotope is being eliminated from the patient.

1. $1 / t_E = 1 / t_H + 1 / t_B$
2. $\lambda_E = \lambda + \lambda_B$ (expressed in terms of the corresponding decay constants)

Radioactive Equilibrium

Radioactive isotopes can decay to form radioactive daughter products that can further decay to form a more stable isotope.

A daughter product will be more stable than its parent if the binding energy per nucleon increases after decay.

If t_H of parent is greater than t_H of daughter, then equilibrium effectively will be achieved after a certain period of time.

Radioactive Daughter

1. Activity of radioactive daughter at time t:

$$A_D(t) = \left[\frac{\lambda_D}{(\lambda_D - \lambda_P)}\right] \cdot A_P(t) \cdot \left[1 - e^{-(\lambda_D - \lambda_P) \cdot t}\right]$$

where

λ_P and λ_D are the decay constants of the parent and daughter, respectively,

and A_P is the activity of the parent.

2. $$\frac{A_D}{A_P} \approx \frac{\lambda_D}{(\lambda_D - \lambda_P)} \approx \frac{t_P}{(t_P - t_D)}$$

where

t_P and t_D are the half-lives of the parent and daughter, respectively.

Metastable State

The excited state of a daughter is said to be in a metastable state if the excited nucleus persists for a considerable amount of time.

Transient Equilibrium

1. Transient equilibrium occurs when the t_H of the parent is not much greater than the t_H of the daughter.
2. Transient equilibrium can be achieved in roughly 4 half-lives of the daughter.
3. Example: ^{99}Mo to ^{99m}Tc

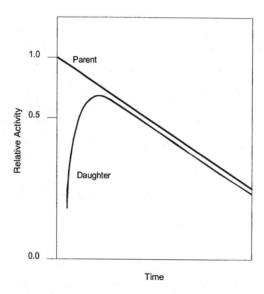

Figure 2.1. Transient Equilibrium

Secular Equilibrium

1. Secular equilibrium occurs when t_H of parent is much greater than t_H of daughter.
2. Example: ^{226}Ra to ^{222}Rn

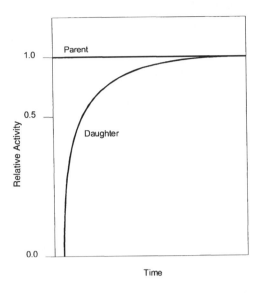

Figure 2.2. Secular Equilibrium

Modes of Radioactive Decay

Nuclei with nucleons in an excited state will "try" to get rid of excess energy (isomeric decay).

Nuclei with too many nucleons for the associated binding energy will "try" to improve the ratio of energy per nucleon by emitting an alpha particle (alpha decay).

Nuclei that do not have a stable ratio of neutrons to protons are in an excited state and will "try" to change neutrons to protons or vice versa (isobaric decay).

Isomeric Decay

Process by which an excited nuclear state is brought to a ground state.

1. Gamma Emission: Electromagnetic radiation is emitted from an excited nucleus.

2. Internal Conversion: An orbital electron acquires the excess energy of the nucleus and is ejected from the atom.

a. Ejected electron will have kinetic energy equal to the energy released by the nucleus minus the binding energy of the ejected electron.

b. Ejected electron leaves atom in an excited state.

c. Probability increases with Z and with lifetime of excited state of the nucleus.

d. Internal Conversion Coefficient:

$$(\alpha) = \frac{\text{\# of conversions}}{\text{\# of gammas emitted}}.$$

Alpha Decay

Process by which an unstable atom in a high-energy state emits a helium particle (two protons and two neutrons) in order to lower its energy state. This process can only occur when the nucleus transfers its excess energy to an alpha particle which then escapes.

1. $^A_Z X \rightarrow {}^{A-4}_{Z-2} Y + {}^4_2 He + Q$

2. Almost all Q is acquired by alpha particle as kinetic energy (KE) and usually leaves with about 5-10 MeV.

3. Occurs mostly in atoms with heavy nuclei (Z > 82).

4. Example: ^{226}Ra to ^{222}Rn

Isobaric Decay

1. Beta Minus: Electron ejected from nucleus changing a neutron into a proton.

    a. $^A_Z X \rightarrow {}^A_{Z+1} Y + \beta^- + \bar{v} + Q$

    b. Electrons ejected with spectrum of energies and anti-neutrino always emitted with the remainder of the energy.

    c. More likely with large neutron to proton ratio.

    d. Common in products of ^{235}U or ^{239}Pu fission.

    e. Examples: ^{60}Co, ^{137}Cs, ^{131}I, ^{192}Ir

2. Beta Plus: Positron emitted from nucleus changing a proton into a neutron.

    a. $^A_Z X \rightarrow {}^A_{Z-1} Y + \beta^+ + v + (Q - 1.02 \ MeV)$

    b. Always competes with electron capture (see *Isobaric Decay, 3. Electron Capture* on page 13).

    c. Only occurs at energies above 1.02 MeV.

    d. Positron emitted with spectrum of energies (minimum 0.511 MeV), neutrino always emitted with the remainder of the energy.

e. Positron annihilates with electron to form photons of 0.511 MeV plus any kinetic energy (KE) traveling in directions usually 180° opposed.

   f. More likely with small neutron-to-proton ratio.

   g. Most common in isotopes produced with high-energy particle accelerators.

   h. Examples: ^{11}C, ^{18}F

3. Electron Capture: Electron (usually in K shell) captured by nucleus changing a proton into a neutron.

   a. $^A_Z X + e^- \rightarrow ^A_{Z-1}Y + v + Q$

   b. Competes with Beta Plus decay (see *Isobaric Decay, 2. Beta Plus* on page 12), but can take place when energy is insufficient for the positron emission.

   c. Neutrino carries off the excess energy at the transition.

   d. Atom left in an unstable state.

   e. Examples: ^{125}I, ^{103}Pd

Isomeric Transition

A nucleus after beta or alpha emission may be in a metastable state, a state where the nucleus is stable for a length of time as a separate isomer, and loses energy through photon emission or internal conversion.

Atomic Physics

Auger Emission

An electron vacancy in an atom's K, L, or M shell can be filled by an electron from a higher shell and the excess energy removed by emission of a different electron known as an Auger electron.

1. Example energy of an Auger electron = $BE_{Kshell} - BE_{Lshell}$ [see *Binding Energy (BE) or Mass Defect* on page 6].

2. Relative probability of emission of characteristic radiation to emission of Auger electron is called the fluorescent yield.

Characteristic Radiation

An electron vacancy in an atom's K, L, or M shell can be filled by an electron from a higher shell and the excess energy removed by the emission of an x-ray (see *Characteristic X-Rays* on page 15).

CHAPTER 3

PRODUCTION OF X-RAYS

X-Ray Production

Electrons interact with atoms in target, producing heat and x-rays.

Additional ionizing electrons are produced (delta rays) and absorbed as heat.

Bremsstrahlung Radiation

Radiation produced when an electron gives up part or all of its energy during a "collision" with a nucleus (the electron is slowed by nucleus) mediated by the action of Coulomb forces.

1. Referred to as "white radiation" or "braking radiation".
2. Photon emitted with a spectrum of energies up to initial electron energy.
3. Probability of occurrence proportional to Z^2 of the target.
4. Efficiency of production $\approx 1 \times 10^{-9} \cdot Z \cdot V$ (higher for higher Z).
5. Most x-ray production above 100 kV is bremsstrahlung radiation.
6. For incident electron energies below 100 keV, bremsstrahlung photons tend to be emitted at right angles to the electron's direction; at higher incident electron energies the emission tends to be in an increasingly more forward direction.
7. The bremsstrahlung radiation produced by electrons with megavoltage energies is peaked in the direction of the electron beam.

Characteristic X-Rays

Radiation produced when electron shell vacancy is created and an outer shell electron fills the vacancy, emitting an x-ray with energy equal to the difference in shell binding energies.

1. X-rays emitted with discrete energies characteristic of atoms in the target and the energies between the shells where the transitions take place (L shell x-rays typically are filtered, but not eliminated, by target).
2. Possible characteristic x-ray energies are found by taking energy differences between the K, L, M, and N shells.
3. Characteristic x-rays can arise from the photoelectric interaction (see *Photoelectric Interaction* on page 38) between bremsstrahlung radiation and the target.
4. Typical diagnostic x-ray beam from tungsten has ~30% characteristic x-rays; typical therapy beam has ~3%.

X-Ray Unit

Basic Circuit Principle

Electrons are "boiled" off a filament (cathode), accelerated across a vacuum tube, and bombarded against a positively charged target (anode).

1. Yield of x-rays depends upon Z^2 of target; amount exiting target depends upon Z^3 to Z^5.
2. Diagnostic unit has line focus and/or rotating anode.
3. Therapy unit has instantaneous energy input less than diagnostic tube, but average energy input over a long period is about 10 times greater.

Anode

1. An anode is a positively charged electrode.
2. It is usually made of tungsten ($Z = 74$) or other high Z material with high melting point.
3. Molybdenum is used for mammography tubes for characteristic x-rays of about 17 keV.
4. 98% to 99.5% of applied energy manifests as heat (rotation of anode and oil used to dissipate heat).
5. Angle of target determines size of x-ray field (targets generally angled at 1° to 45° for diagnostic units).

Cathode

1. A cathode is a negatively charged electrode.
2. It consists of a wire filament heated until electrons "boiled" off (thermionic emission).
3. It is usually made of tungsten because of its high resistance (heat) and high melting point.
4. Voltage applied across tube must be sufficient to overcome the space-charge effect (cloud of electrons that forms around a heated filament and repels electrons back into the filament).

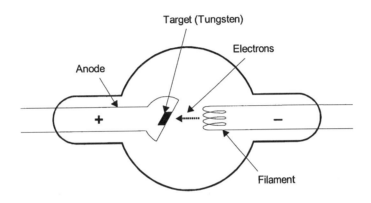

Figure 3.1. Basic X-ray Unit Diagram

Heat Production of Diagnostic Tubes

Heat Units (HUs)

1. Single-Phase: $HU(1\phi) = mA \cdot kV \cdot s$
2. Three-Phase: $HU(3\phi) = 1.35 \cdot mA \cdot kV \cdot s$

Energy Deposited

$E_d(1\phi \text{ or } 3\phi) \approx 0.75 \cdot HU$

Focal Spot

Focal spot size increases in direct proportion to tube current, and decreases slightly with increasing kVp.

Projected length of focal spot is much shorter at the anode side than at the cathode side.

X-ray beam intensity in the anode-cathode direction is greater toward the cathode side (heel effect) with approximately 30% reduction in useful beam on the anode side (depending on the angle of the x-ray target).

X-Ray Tube Operating Characteristics

Output linear with mA.

Output proportional to the square or higher of kV (greater increase at higher energies).

Output very dependent on filament current (more than kV at about 75 kVp).

Output increases with Z of target (greater increase at higher energies).

Output reduced and effective energy increased with filtration.

Filters

Filters are added to x-ray beams to preferentially absorb the lowest energy photons, reducing beam intensity and increasing beam hardening.

Inherent Filtration

1. Inherent filtration includes the filtration created by attenuation from the glass envelope of tube, surrounding oil, and exit window.
2. Inherent filtration is roughly equivalent to 1 mm of aluminum for most diagnostic x-ray tubes.

Diagnostic

1. Usually aluminum filtration is added to provide at least a half-value layer (HVL) [see *Half-Value Layer (HVL or X_H)* on page 37] of 2.5 mm at a peak photon energy of 80 kVp.
2. The filtration eliminates radiation of very low energy that would likely be absorbed in a patient and not reach an image receptor.

Therapy - Graded (Thoraeus)

1. X-ray beam passes through layers of decreasing Z material (tin, copper, and then aluminum).
2. Used to increase HVL of orthovoltage beam without reducing beam intensity to an unacceptable level.
3. Low Z material of filter is used to absorb characteristic x-rays produced by photoelectric effect in higher Z material.

Energy Spectra

An x-ray unit's energy spectrum comprises a continuous spectrum of bremsstrahlung with characteristic radiation spikes.

"Thin Target" Approximation

1. Thin target approximation assumes no electron suffers more than one collision on average while passing through the target.
2. Energy radiated is simply: $I(E) = C \cdot Z \cdot E$

    where

    I is the intensity of the photons with energy E,

    Z is the atomic number of the target, and

    C is a constant.

"Thick Target" Approximation

1. Energy radiated (assuming no filtration spectrum) is given by Kramer's equation: $I(E) = C \cdot Z \cdot (E_{max} - E)$

    where

    E_{max} is the maximum energy radiated.

2. If filtration is considered: $E_{avg}(h\nu) \approx 1/3 \cdot \text{max kVp}$.

X-Ray Image

Image Quality

1. Grids
    a. A series of lead strips separated by spacers transparent to x-rays used to discriminate against scattered radiation.
    b. Grid ratio $(GR) = h/b$

    where

    h is the height of the high density strips, and

    b is the width of the low density material spacers.

2. Quantum mottle: An important factor affecting image quality and caused by the statistical fluctuation in the number of x-ray photons per unit area absorbed by a receptor.

Modulation Transfer Function (MTF)

Evaluates the ability of an imaging system to record an image by quantifying how well the system captures contrast at spatial frequencies.

1. $MTF = \dfrac{\text{(modulation of output signal)}}{\text{(modulation of input signal)}}$.

2. Experimentally determined for a x-ray imager with a square wave (bar) test pattern, using Fourier transforms to convert output to a sinusoidal pattern to match the alternating current (sinusoidal) input.

Computed Tomography (CT)

CT Number

CT Number = $K \cdot (\mu_{tissue} - \mu_{water}) / \mu_{water}$

where

> K is a manufacturer-dependent magnification constant and
>
> μ is the linear attenuation coefficient for the pixel (see *Exponential Attenuation* on page 35).

Hounsfield Unit (HU)

Hounsfield Unit (HU) = $1000 \cdot (\mu_{tissue} - \mu_{water}) / \mu_{water}$

1. $HU_{water} = 0$
2. $HU_{air} \approx -1000$
3. $HU_{bone} \approx +1000$

Mammography

Tube potential is typically below 30 kVp because structures of low contrast dictate that a low energy spectrum should be used.

A molybdenum anode with a 30 μm molybdenum filter provides a beam with a large proportion of 17.5 keV photons, which provides good contrast between fat, muscle and calcifications.

CHAPTER 4
THERAPY RADIATION GENERATORS

Kilovoltage Units

Grenz

1. <20 kVp
2. Filtration of ~0.1 mm Al.
3. Used by dermatologists to treat skin lesions.

Contact Therapy

1. 40 – 50 kVp
2. Filtration of 0.5 – 1 mm Al.
3. Typical source-to-surface distance (SSD) of about 5 cm.
4. Dose falls off rapidly with depth.

Superficial

1. 50 – 150 kV, 5 – 8 mA
2. Filtration of 1 – 10 mm Al.
3. Typical SSD of 15 – 20 cm.
4. Used to treat superficial lesions.

Orthovoltage

1. 150 – 500 kV, 10 – 20 mA
2. Filtration of 1 – 4 mm Cu.
3. Peakscatter factor (PSF) up to 1.5 (largest found with x and gamma rays) [see *Peakscatter Factor (PSF)* on page 66].
4. Greatest limitation to treatment of deep targets is high skin dose.

Megavoltage Units

Van de Graaff

Electrostatic accelerator of electrons up to 3 MeV.

Betatron

Electrons accelerated in a changing magnetic field (magnetic induction) through a circular orbit.
1. Electrons accelerated from 6 to 45 MeV.
2. Limited by low photon dose rate and large size.
3. Has advantage of higher fractional depth dose (FDD) [see *Fractional Depth Dose (FDD)* on page 64] than a linear accelerator for a nominal energy because of the use of a thin target.

Cyclotron

Charged particle accelerator based on the principle of alternating the polarity of evacuated dees timed with the circular flight of accelerating charged particles.
1. Good for producing beams of protons (and, through interactions, neutrons), but not used to accelerate electrons.
2. Often used to produce short-lived β^+ emitters such as ^{15}O.

Microtron

Electrons are accelerated by an oscillating electric field in a microwave cavity and forced by a magnetic field to move in a circular orbit and then return to the cavity. Repeated passes through the cavity increase the energy of the electrons and the radius of the path they travel until they reach an appropriate path and are extracted from their orbit.

Linear Accelerator (LINAC)

Charged particles are accelerated to high energies through a linear tube by high-frequency electromagnetic waves.

1. Traveling Wave
   a. High-frequency microwaves of 3000 MHz are transmitted down an evacuated tube through evenly spaced accelerating cavities $\lambda/4$ (~2.5 cm) in length.
   b. A prebuncher is used to reduce velocity of the electromagnetic wave in order to correspond to the speed of the injected electron so that the electron remains on the crest of the wave and undergoes acceleration.
   c. Energy transfers of ~75 keV / cm are possible.
   d. The electromagnetic waves are absorbed in a dummy load at the end of the guide to prevent them from reflecting and interfering with incoming waves.

2. Standing Wave
   a. A standing wave is produced when two traveling waves of equal amplitude and period travel through a waveguide in opposite directions.
   b. The coupling of adjacent cavities enables an electron to experience acceleration through each accelerating cavity.
   c. Energy transfers of ~150 keV / cm are possible.
   d. Standing wave accelerator structures are used in most modern LINACs.

3. Components
   a. Electron gun: Cathode that provides source of bunched or pulsed electrons injected into accelerator structure.
   b. Magnetron: Produces microwaves with $\nu = 3000$ MHz, originally used for radar and microwave ovens, operates at about 2 MW peak output, typically used to produce electron beams up to ~15 MeV.
   c. Klystron: Microwave amplifier, driven by a low-power microwave oscillator, operates at peak output of 5 MW.
   d. Modulator: Simultaneously provides high voltage direct current pulses to the magnetron or klystron and the electron gun.
   e. Waveguide: Carries microwave power from a magnetron or klystron through accelerator structure.
   f. Accelerator structure: Accelerates electrons from an electron gun using microwave power from a magnetron or klystron.

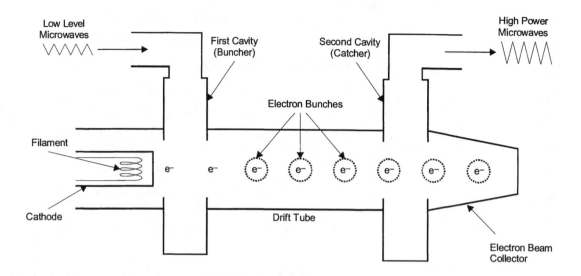

Figure 4.1. Klystron Schematic

4. Treatment Head

   a. Bending magnet: Deflects electrons from accelerator structure and loops the electrons around to strike the target for photon production or scattering foil for electron beam treatments.

   b. Scattering foil: Thin metallic foil used to broaden electron beam to a useful-sized field (some LINACs broaden the electron beam by electromagnetically scanning the beam over an area).

   c. Target: High Z material (usually tungsten) used to produce x-rays through bremsstrahlung interactions (see *Bremsstrahlung Radiation* on page 15).

   d. Flattening filter: Material such as lead, tungsten, uranium, aluminum, brass or some combination of these used to make photon beam intensity more uniform.

Cobalt-60 Units

Source

1. ^{60}Co is produced using neutron bombardment of ^{59}Co in nuclear reactors.
2. Source diameter is typically between 1 and 2 cm.

3. Emits two useful gammas of energy, 1.17 and 1.33 MeV (average = 1.25 MeV), depth of maximum dose in water ≈ 0.5 cm for a 10 cm × 10 cm field [see *Depth of Maximum Dose* (d_{max}) on page 61].

4. Half-life = 5.26 years [see *Half-Life* (t_H) on page 8] so source decays at about 1% / month (decay correction = 0.9891).

Dosimetric Characteristics

1. Half-value layer (HVL) = 1.1 cm of lead [see *Half-Value Layer* (*HVL or* X_H) on page 37].

2. Attenuation ≈ 4% – 5% / cm in tissue (similar fractional depth dose (FDD) to 14 MeV neutron).

3. Γ = 13.07 R · cm² / mCi · hr (see *Exposure Rate* on page 78).

CHAPTER 5

INTERACTIONS OF ELECTRONS AND CHARGED PARTICLES WITH MATTER

Ionization and Excitation

Ionization

The removal of an orbital electron from an atom that then acquires a charge.

1. Direct ionization is produced when charged particles have sufficient kinetic energy (KE) to cause ionization and therefore deliver their energy directly to matter.
2. Indirect ionization is produced by uncharged particles such as neutrons and photons (see Chapter 6) that cause the liberation of directly ionizing particles.

Excitation

The displacing of an orbital electron from its ground state followed by a return after a transfer of energy sufficient to overcome the binding energy.

Collisions

Interactions (electron energy losses) with tissue are principally by ionization and excitation.

Interactions are mediated by Coulomb forces between the electric field of the traveling particle and electric fields of orbital electrons and nuclei of the atoms of a material.

Momentum is always conserved for a collision.

Total Energy is always conserved for a collision, but kinetic energy (KE) is not.

Elastic Collisions

1. Predominantly ionizing events with atomic electrons where no KE is lost.
2. Involves light particles such as electrons.

Inelastic Collisions

1. Predominantly ionizing events with atomic electrons that result in a loss of KE.
2. The KE lost can be converted into photon energy or excitation energy.
3. Involves light and heavy particles.

Energy Losses

All charged particles lose kinetic energy chiefly through interactions between the electric field of the particle and electric fields of electrons in the material through which the particle is traveling.

Rate of energy loss for charged particles is proportional to the square of the particle's charge and inversely proportional to the particle's velocity.

Rate of total energy loss per cm is approximately proportional to the energy (E) of the particle and to the square of the medium's Z.

Probability of radiation energy loss relative to collisional energy loss increases with E and Z.

Scattering power varies approximately as the square of Z and inversely as the square of the KE; therefore, scattering foils are made with high Z materials and are thin to minimize x-ray contamination [see *Linear Accelerator (LINAC), 4. Treatment Head* on page 26].

Energy loss rate of electrons with greater than 1 MeV is approximately 2 MeV / cm.

Stopping Power (S)

Rate of energy loss per unit path length by a charged particle traveling through a material.

1. Total losses can be separated into collisional and radiative components:

    $(S/\rho)_{tot} = (S/\rho)_{col} + (S/\rho)_{rad}$

    where

    > ρ is the density of the material traversed and

    > S/ρ represents the mass stopping power.

2. $(S/\rho) = (1/\rho) \cdot (dE/dx)$

    where

    > dE/dx is the energy loss per centimeter and

    > ρ is expressed in g / cm^3 so that (S/ρ) is expressed in (MeV $\cdot$ cm^2 / g) or in SI units (J $\cdot$ cm^2 / g).

3. Stopping power is greater for low Z material than for high Z because high Z materials have fewer electrons per gram and more tightly bound inner electrons.

Restricted Stopping Power (L)

Restricted stopping power is similar to stopping power (S) but excludes energy carried away by delta rays (ejected electrons with enough energy to cause secondary ionization).

1. Restricted stopping power (L) considers collisional losses only and only collisions with secondary electrons with less than some delta (Δ) energy cutoff (usually 10 keV).

2. Restricted mass stopping power (L/ρ) is the restricted stopping power per density.

3. Used with Bragg-Gray cavity theory for ionization chambers (see *Bragg-Gray Cavity Theory* on page 52).

Range

The finite distance beyond which conceptually there will be no particles. As a beam of charged particles passes through matter, the particles continually lose kinetic energy in interactions causing them to slow down and change directions, and at some point they run out of energy and stop.

1. Practical Range (R_p) or Extrapolated Range is the intersection of the extrapolation of the straight descending portion of the fractional depth dose (FDD) curve and the extrapolation of the bremsstrahlung background (R_p for electrons is approximately E / 2 MeV/cm where E is the incident electron energy).
2. CSDA Range or Continuous Slowing Down Approximation Range is calculated from determining energy loss using mass stopping power, taking into account the variation of $(S/\rho)_{tot}$, in a material with constant density ρ with respect to energy.

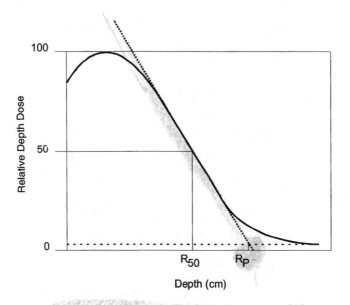

Figure 5.1. Electron Depth Dose Curve

Bragg Peak

Bragg Peak is an area near the end of a track of a charged particle traveling through a medium where the rate of energy loss increases very rapidly and then falls to nearly zero.

1. A Bragg Peak is observed for heavy charged particles because the rate of energy loss (S) increases as the particles' velocity decreases and because heavy particles suffer little multiple or side scattering.
2. An electron beam slows down with multiple changes in direction, causing the Bragg peak to smear out and not be observed.

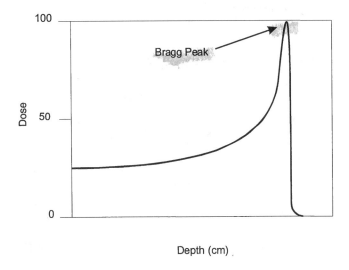

Figure 5.2. Heavy Charged Particle Depth Dose Curve

Bremsstrahlung Yield (B)

Bremsstrahlung Yield (B) = $\dfrac{1}{E_0} \cdot \int_0^{E_0} \dfrac{S_{rad}(E)}{S_{tot}(E)} dE$.

where

E_0 is the initial energy of the electron,

S_{rad} is the energy loss by radiation as the electron travels through the medium, and

S_{tot} is the total energy loss of the electron.

Polarization ("Density") Effect

Electrons in intervening atoms can influence interactions with more distant electrons.

The intervening atoms become polarized by the electric field. This polarization reduces the effect of the electric field of the incident charged particle at distant points.

Electron Energy Spectrum

Energy of electrons decreases monotonically with depth in a phantom.

A Gaussian distribution can approximate the angular and spatial spread of a narrow, collimated beam of electrons.

Most Probable Incident Energy

1. Nordic Association of Clinical Physics relationship:

    E_p (MeV) = $0.22 + (1.98 \cdot R_p) + (0.0025 \cdot R_p^2)$

    where

    the field size is at least 12 cm × 12 cm for electron energies up to 10 MeV

    and 20 cm × 20 cm for energies greater than 10 MeV.

2. Markus relationship: E_p (MeV) = $0.722 + (1.919 \cdot R_p)$.

Mean Incident Energy

$(E_0) = 2.33 \cdot R_{50}$

where

R_{50} is the depth at which the dose is 50% of the maximum dose.

Energy at Depth

$(E_p)_z = (E_p)_0 \cdot [1 - (d / R_p)]$

where

$(E_p)_0$ is the most probable energy at the surface and

d is the depth.

CHAPTER 6

INTERACTION OF IONIZING RADIATION WITH MATTER

Interaction of Photon Beam

X-ray and gamma-ray photons interact with matter by absorption and scatter.

Absorption gives rise to electrons that travel through matter producing ionization and excitation (see *Ionization and Excitation* on page 29) and transfer energy.

Scatter gives rise to secondary, lower-energy photons.

Description

1. Fluence: $\Phi\ [m^{-2}] = dN/da$

    where

    N is the number of photons and

    a is the area considered.

2. Fluence Rate (flux density): $\phi\ [m^{-2}s^{-1}] = d\Phi/dt$.

3. Energy Fluence: $\Psi\ [Jm^{-2}] = dE_f/da$

    where

    E_f is the sum of energy of all photons.

4. Energy Fluence Rate (energy flux density): $\psi\ [Wm^{-2}] = d\Psi/dt$.

Exponential Attenuation

1. Photon beam attenuation is described by: $I(x) = I_0 \cdot e^{-\mu x}$

    (valid when μ is constant, which is true for a narrow, monoenergetic photon beam),

    where

μ is the linear attenuation coefficient and represents the fraction of photons that interact per unit thickness of attenuator.

2. Attenuation typically depends on photon energy and the nature of the material (atomic number and density).
3. Linear attenuation represents the fractional change in fluence per unit distance through a medium.
4. Mean path length is the average distance traveled by photons before interaction in a particular medium and is equal to μ^{-1}.

Attenuation Coefficients

1. Mass attenuation coefficient: $(\mu/\rho) = \mu/\rho$
   where
   ρ is density of the attenuator.
   a. The attenuation coefficient depends on the atomic composition of the material rather than the density.
   b. Mass attenuation coefficient for compound:
   $(\mu/\rho)_{mix} = (\mu/\rho)_A \cdot F_A + (\mu/\rho)_B \cdot F_B + ...$
   where
   F_i is the weighted fraction of component i.

2. Total mass attenuation coefficient: $(\mu_{tot}/\rho) = (\sigma_C/\rho) + (\tau/\rho) + (\sigma_I/\rho) + (\kappa/\rho)$
   where the attenuation coefficients have been separated into the various interaction processes that can occur. These are:
   a. Coherent scattering (σ_C) (see *Coherent Scattering* on page 37).
   b. Photoelectric process (τ) (see *Photoelectric Interaction* on page 38).
   c. Incoherent or Compton scattering (σ_I) (see *Incoherent or Compton Scattering* on page 39).
   d. Pair production (κ) (see *Pair Production* on page 42).

Energy Coefficients

1. Mass energy transfer coefficient: $(\mu_{tr}/\rho) = (\mu/\rho) \cdot (E_{tr}/h\nu)$
   where
   E_{tr} is the average energy transferred to kinetic energy of charged particles per photon ($h\nu$) interaction (see *Kerma* on page 51).

2. Mass energy absorption coefficient: $(\mu_{ab}/\rho) = (\mu/\rho) \cdot (E_{ab}/h\nu) = (\mu_{tr}/\rho) \cdot (1-g)$

   where

   E_{ab} is the average energy absorbed in the medium at the point of interaction (see *Absorbed Dose* on page 51) and

   g represents the fraction of radiative energy loss in the material.

Half-Value Layer (HVL or X_H)

1. HVL is the thickness of a given material required to reduce a beam to half its original value.
2. $X_H = 0.693 / \mu = \ln 2 / \mu$.
3. HVL is dependent on waveform and kVp, but independent of mA for x-ray beams.
4. Lower-energy (softer) components of a photon spectrum are attenuated at a quicker rate than the higher-energy (harder) components by a material, resulting in beam hardening and an increase in HVL with thickness.

Spectrum

1. Mean energy: Average energy defined by the spectral distribution of a radiation field and characterized by the distribution of fluence or energy fluence with respect to energy.
2. Effective energy: Energy of photons in a monoenergetic beam which is attenuated at the same rate as the radiation in question.

Coherent Scattering

During coherent scattering no energy of a "colliding" photon is converted to kinetic energy; instead, all energy is scattered in a mostly forward direction, broadening the angular width of the beam.

1. Effect is significant for interactions of photons with very low energies (<10 keV).
2. Coherent scattering accounts for roughly 5% of total scatter from a diagnostic x-ray beam.

3. Probability of coherent scattering (σ_C/ρ) increases with decreasing photon energy and increasing Z of the material.
4. Coherent scattering is neglected in radiotherapy because no energy is absorbed. Since the scattering is predominantly in the forward direction, the coherent events do not alter the properties of high-energy beams significantly.

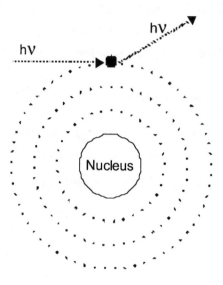

Figure 6.1. Coherent Scattering

Thomson (Classical)

The electric field of a photon passing near an electron can cause it to be momentarily accelerated and radiate energy.

Rayleigh (Coherent)

The combination of photon scattering with bound electrons from different parts of the atomic cloud in a process where the atom is left neither excited nor ionized.

Photoelectric Interaction

Photoelectric interaction occurs when photon energy is transferred to an atom and a bound electron (photoelectron) is ejected.
1. Involves mostly inner shell bound electrons, but can take place with electrons in the K, L, M or N shells.

2. Probability of electron ejection is a maximum if the photon has just enough energy to knock the electron from its shell.

3. Photoelectrons have kinetic energy equal to the difference between the incident photon energy and the binding energy of the ejected electron (rarely enough energy to travel a centimeter in tissue).

4. A more outer shell electron fills the vacancy left by the removal of an electron giving rise to Auger electrons (see *Auger Emission* on page 13) and characteristic x-rays (see *Characteristic X-Rays* on page 15).

5. The photoelectric attenuation coefficient (τ/ρ) is approximately proportional to E^{-3}, Z^3 for high Z materials, and $Z^{3.5}$ to Z^5 for low Z materials.

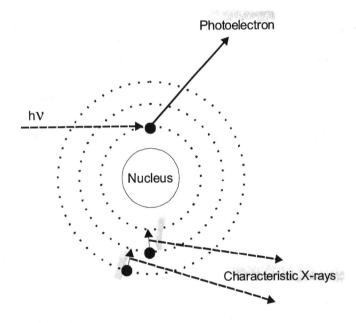

Figure 6.2. Photoelectric Effect

Incoherent or Compton Scattering

Incoherent or Compton scattering involves a photon collision with a "free" or loosely bound electron (appearing free to a high-energy photon) where the photon energy is transferred to an ejected Compton electron and to a lower-energy scattered photon. High-energy photons (10 to 100 MeV) give most of their energy to the Compton electron, while low-energy photons give most of their energy to the scattered photon.

1. Incoherent or Compton scattering is dependent on electron density, but almost independent of Z when Compton interactions measured per gram per cm² as opposed to per cm (ratio of Compton interactions in 1g of hydrogen to 1 g of another atom is ≈ 2 due to electrons/g which is nearly equal for all elements except hydrogen).
2. Compton attenuation coefficient (σ_r/ρ) decreases with an increase in photon energy and is independent of the Z of a material.
3. The Compton process is the most important interaction mechanism in tissue-like materials.

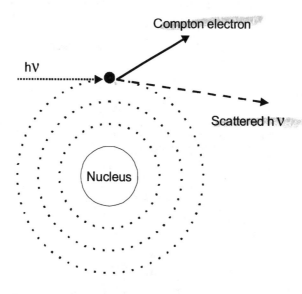

Figure 6.3. Compton Effect

Energy-Angle Relationships

1. Compton electron energy: $(E_C) = E_{hv} \cdot \left\{ \dfrac{[\alpha \cdot (1-\cos\phi)]}{[1+\alpha \cdot (1-\cos\phi)]} \right\}$

   where

   E_{hv} is the incident photon energy,

   α is the ratio of E_{hv} to the rest energy of an electron (0.511 MeV), and

   ϕ is the angle the scattered photon makes with respect to the incident photon.

2. Compton electron angle of emission $(\theta) = \cos^{-1}[(1 + \alpha) \cdot \tan(\phi/2)]$

   where

   θ is the angle the ejected electron makes with respect to the incident photon.

3. Scattered Photon Energy: $(E_{h\nu'}) = E_{h\nu} \cdot [1 / (1 + \alpha \cdot (1 - \cos\phi))]$.

4. Examples of possible Compton Interactions:
   a. Grazing hit ($\theta = 90°$, $\phi \approx 0°$): The electron will emerge at nearly a right angle and the scattered photon will go almost straight forward with an energy equal to the incident photon.
   b. Direct hit ($\theta = 0°$, $\phi = 180°$): The electron will travel straight forward with maximum energy and the scattered photon will be scattered straight back with minimum energy (backscattered photon has maximum energy of ≈ 256 keV).
   c. Side-scatter photon ($\phi = 90°$): The scattered photon will emerge with a maximum energy of 511 keV and the electron energy and angle will depend on the incident photon energy.

Klein-Nishina Coefficients

1. The probability of Compton collision (total Compton coefficient) with a "free" electron may be determined by the Klein-Nishina cross-section determined using suitable wave functions and quantum mechanics.
2. Energy transfer and scatter coefficients for Compton interactions can be determined from the total Compton coefficient and the fraction of energy given to the Compton electron.

Pair Production

Pair production involves an interaction between a photon and a nucleus where the energy of the photon is transformed into an electron and a positron. Two 0.511 MeV annihilation photons are radiated from the absorber when the positron later interacts with an electron.

1. The threshold for the pair production process is 1.02 MeV, thereafter, the energy transferred to kinetic energy (KE) is: $E_{tr} = E_{hv} - 1.02$ MeV

   where

   E_{hv} is the incident photon energy.

2. The electron and positron energy distribution is generally split approximately equally.

3. The probability of a pair production interaction (κ/ρ) increases rapidly with energy above threshold (with the logarithm of incident photon energy), and varies approximately as Z^2 per atom, as Z per electron, and as Z per gram.

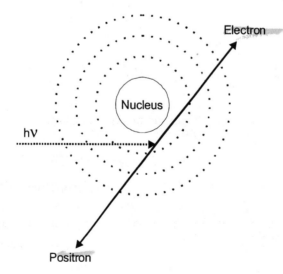

Figure 6.4. Pair Production

If a pair production interaction occurs in the vicinity of an electron instead of at a nucleus, triplet production occurs with two electrons and one positron leaving the site of interaction. The threshold for triplet production requires 2.04 MeV to conserve momentum and energy.

Photodisintegration

Photodisintegration occurs when a photon interacts with and is captured by a nucleus. The nucleus becomes unstable and has the potential to release nucleons (mostly neutrons) or gamma rays.

Relative Importance of Interactions

Coherent scattering is significant for less than 10 keV photons and in high Z materials.

Photoelectric absorption is important for photon energies of 10 keV to 1 MeV and in high Z materials, and is responsible for beam hardening [see *Half-Value Layer* (*HVL* or X_H) on page 37] in flattening filters [see *Linear Accelerator (LINAC)*, *4.d. Flattening Filter* on page 26] of photon beams up to 10 MeV.

Compton scattering is important for photon energies from 30 keV to 10 MeV and is independent of the Z of the material.

Pair production is observed with photons above 1.02 MeV and is important in high Z materials up to 100 MeV.

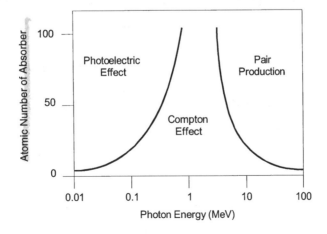

Figure 6.5. Relative Importance of Principal Interactions

CHAPTER 7

MEASUREMENT OF IONIZING RADIATION AND ABSORBED DOSE

The Roentgen

The roentgen is the unit of exposure and a measure of the ability of a radiation to ionize air. It is defined as the absolute value of the total charge of the ions of one sign produced in air when all of the electrons liberated by photons in a volume element of air having a specified mass are completely stopped in air.

1. It is defined only for photons.
2. It is difficult to measure for photons >3 MeV because of the thickness of air required for electronic equilibrium [see *Electronic Equilibrium* (*EE*) on page 49].

Exposure (X) = dQ/dm

1. 1 R = 1 electrostatic unit (esu) of charge per cm^3 of dry air at standard temperature and pressure or 3.33×10^{-10} C (see *Temperature and Pressure* on page 5).
2. R = 2.58×10^{-4} C/kg of air.

f-factor (f)

The f-factor converts exposure to dose and is a function of the absorbing medium and the energy of the radiation.

1. In air, f is taken as 0.876 cGy/R (or rad/R) for low energy photons.
2. For other materials: $f_{med} = 0.876 \cdot \left(\dfrac{\bar{\mu}_{en}}{\rho}\right)^{med}_{air}$

3. f for various photon energies in water is:

Energy (MeV)	f - factor
0.12	0.89
^{137}Cs (0.662)	0.95
^{60}Co (1.25)	0.96
4	0.94
6	0.94
18	0.91

Chambers

Chamber Types

1. Free-Air Ionization Chamber: Instrument employed in the absolute measurement of the roentgen according to its definition.
   a. Beam axis is parallel to the collection volume.
   b. Designed such that electrons liberated outside the collection volume but collected by the electrodes are balanced by those liberated inside the volume and not collected.
   c. Range of electron increases with photon energy so useful energy range is limited by size of chamber.

2. Parallel-Plate or Plane-Parallel Chamber: Chamber where voltage is applied to two thin and closely separated parallel plates that collect ions.
   a. Beam axis is usually perpendicular to the electrode plates.
   b. Measurement position is taken as the inside of the front electrode.
   c. Chamber recommended for measurement of electron beams with E < 10 MeV.

3. Extrapolation Chamber: A parallel-plate ionization chamber used for measuring surface dose in a medium.
   a. The spacing of the electrodes can be varied accurately.
   b. Ionization at the surface of a medium determined by plotting dI/dx versus x, where I is the measured current and x is the plate separation, and extrapolating to zero electrode spacing.

4. Thimble Ionization Chamber: A coaxial-designed chamber with one electrode forming a thimble-shaped shell around the collecting volume and the other electrode as a narrow central rod. The Farmer chamber is one example of a thimble chamber that is often used to measure charge produced by high-energy radiation beams.

Chamber Wall

1. For lower-energy photons (<1 MeV), most ionization produced in the collecting volume of a chamber's cavity arises from electrons liberated in the surrounding wall.
2. The wall thickness must be sufficient to achieve electronic equilibrium [see *Electronic Equilibrium* (*EE*) on page 49], but thin enough to not significantly attenuate the photon flux.
3. For higher-energy photons (radiations from ^{60}Co and photons >0.5 MeV), a build-up cap may be used over the sensitive volume of the chamber with a sufficient thickness to provide the required equilibrium.
4. A correction factor to account for the attenuation of the radiation beam by the chamber wall can be obtained by extrapolating linearly the attenuation curve normalized to the maximum response (Figure 7.1) beyond the maximum and back to a zero thickness.

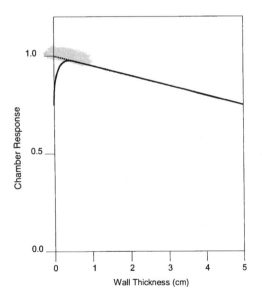

Figure 7.1. Chamber Response Versus Wall Thickness

Stem Effect

1. The stem effect is caused by the contribution of ionization measured by a chamber system that was produced outside the chamber proper in the body of the stem.
2. A stem correction factor can be determined as a function of stem length exposed relative to the amount of the stem exposed under calibration conditions.

Desirable Chamber Characteristics

1. Minimal variation with radiation energy.
2. Minimal variation over range of reasonable exposures.
3. Minimal perturbation of measured radiation beam.
4. Minimal stem leakage and ion recombination losses.

Ion Collection

Saturation

1. The ionization current measured by an ion chamber exposed to radiation first increases with an increase in the voltage difference between the electrodes, begins to level off, and then reaches a saturation value for a particular exposure rate.
2. An ion chamber should be used in the saturation region so that the voltage is large enough to overcome ion recombination, but low enough that accelerated ions do not gain enough energy to produce ionization by collision with gas molecules in the chamber.

Recombination

Some negative and positive ions in a collecting volume still tend to recombine when collected by an appropriate electric field, causing a loss of charge collected that must be corrected for.

1. The recombination correction factor for pulsed or continuous radiation can be determined by measuring the ionization at two bias voltages (one twice the voltage of the other) and taking their ratio.

2. Recombination is more likely to occur with high LET (linear energy transfer) [see *Linear Energy Transfer (LET)* on page 88] radiation, greater dose rate, increased electrode spacing, decreased polarizing voltage, and increased gas temperature.

Temperature and Pressure

1. If an ion chamber is not sealed to the outside atmosphere, the air temperature and pressure (atmospheric density) will affect its response at the time of measurement.
2. If the initial chamber calibration is normalized to the atmospheric conditions of 22 °C and 760 mmHg, then the temperature (T) and pressure (P) correction factor at the time of measurement is: $C_{TP} = \left[\frac{(273.15\,°C + T)}{(295.15\,°C)}\right] \times \left(\frac{760\,mmHg}{P}\right)$.

W/e Factor

The W/e factor represents the average energy required to cause one ionizing event in the gas of a measuring cavity.

1. W/e has units of eV per ion pair (eV/IP) or equivalently J/kg.
2. W/e for air has a nearly constant value of 33.97 eV/IP for all electron energies.

Electronic Equilibrium (EE)

Electronic equilibrium (EE) exists when the ionization lost from a measuring volume is compensated by the ionization entering that volume.

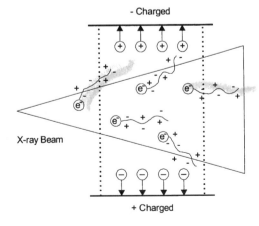

Figure 7.2. **Electronic Equilibrium**

1. Complete electronic equilibrium does not truly exist for high-energy photon beams.
2. Typical EE depths in water for various photon energies:

Energy (MeV)	EE Depth (cm)
^{137}Cs (0.662)	0.3
^{60}Co (1.25)	0.5
4	1.0
6	1.5
18	3.0

Polarity Effects

The ionic charge collected by an ion chamber for a given exposure can sometimes change as the polarity of the collecting voltage is reversed.

1. Causes of polarity effects include Compton current and extracameral current.
   a. Compton current occurs when high-energy electrons ejected by high-energy photons create a current independent of gas ionization that may add to or reduce the collector current, depending on the polarity of the collecting electrode.
   b. Extracameral current occurs when Compton current, or current generated by some other source such as when the connecting cable is irradiated, is collected outside the sensitive volume of the chamber.
2. Polarity effects are generally more severe for measurements made in electron beams than in photon beams and the effect increases with decreasing electron energy.
3. Polarity corrections are made by applying a correction factor P_{pol} that is determined from uncorrected ion chamber readings taken under reference dosimetry conditions at positive (M^+) and negative (M^-) collecting voltages and applying the equation:

$$P_{pol} = \frac{(M^+ - M^-)}{[2 \cdot (M^+ \text{ or } M^-)]}$$

Energy Transfer

Kerma (Kinetic Energy Released in the Medium)

The transfer of energy from uncharged particles (such as photons) to a medium through setting charged particles in motion.

1. Kerma [J/kg] = dE_{tr}/dm (kinetic energy transferred to charged particles per unit mass).
2. Kerma can be divided into an inelastic component (see *Inelastic Collisions* on page 30) and a radiative component (see *X-Ray Production* on page 15).
3. Collision Kerma in air is directly proportional to exposure, the proportionality constant being equal to the W/e factor (see *W/e Factor* on page 49).

Absorbed Dose

The energy actually absorbed in a medium from impinging radiation (energy not radiated away) through ionization and excitation processes that take place (see *Ionization and Excitation* on page 29).

1. Absorbed dose [J/kg] = dE_{ab}/dm (energy absorbed by a medium per unit mass).
2. 1 Gray = 100 cGy = 1 J/kg = 100 rad where 1 rad represents the absorption of 100 ergs of energy per gram of absorbing material.
3. Energy lost through radiative processes from the secondary electrons is small beyond the depth of electronic equilibrium: $E_{tr} \approx E_{ab}$ and $\mu_{tr} \approx \mu_{ab}$ (see *Energy Coefficients* on page 36) in low Z material.
4. Absorbed dose is roughly a measure of the biological effects from ionizing radiation exposure when all other parameters are held constant.

Energy Transfer and Dose Curves

1. The Kerma curve has a maximum value at surface and then continually decreases with depth.
2. For the absorbed dose curve, dose increases through an electron build-up region and then decreases somewhat linearly (without scatter the decrease would be more exponential).

3. The area under the two curves taken to infinity must be equal after taking into account any radiation loss.

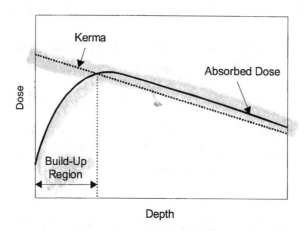

Figure 7.3. Absorbed Dose Versus Kerma

Bragg-Gray Cavity Theory

Bragg-Gray cavity theory relates the ionization produced in a gas-filled cavity to the energy absorbed in the surrounding medium.

1. Not subject to the limitations of calculating absorbed dose from exposure (can be used for photons >3 MeV and with charged particles).
2. Assumes that the cavity is small enough that it does not perturb the particle fluence that would exist in the medium without the cavity.

Bragg-Gray Relationship

$$D_{med} = J_{gas} \cdot (W/e)_{gas} \cdot (S/\rho)_{gas}^{med}$$

1. J_{gas} (Q/m_{gas}) is the ionization charge of one sign produced per unit mass of cavity gas.
2. Refer to *W/e Factor* on page 49 for a discussion of (W/e).
3. $(S/\rho)_{gas}^{med}$ is the mean ratio of the mass stopping power of the medium to that for the cavity gas for electrons crossing the cavity [see *Stopping Power (S)* on page 31].

Spencer-Attix Formulation of Bragg-Gray Relationship

$$D_{med} = J_{gas} \cdot (W/e)_{gas} \cdot (L/\rho)_{gas}^{med}$$

1. $(L/\rho)_{gas}^{med}$ is the mean ratio of the restricted mass collision stopping power of the medium to that for the cavity gas with a Δ cutoff energy [see *Restricted Stopping Power (L)* on page 31] whose value is approximately the energy of an electron that will just cross the cavity.
2. The Spencer-Attix Formulation accounts for the effects of secondary electrons produced.

Absorbed Dose in a Medium (TG-21 Calibration Protocol)

Absorbed Dose from Photons

$$D_{water} = M \cdot N_{gas} \cdot (L/\rho)_{air}^{med} \cdot P_{ion} \cdot P_{repl} \cdot P_{wall} \cdot (\mu_{ab}/\rho)_{med}^{water} \cdot SC$$

1. $M = M_{raw} \cdot C_{TP} \cdot P_{elec} \cdot P_{pol}$, which is the charge collected in a measuring chamber in coulombs corrected for temperature, pressure, electrometer inaccuracies, and polarity effects.
    a. M_{raw} is the uncorrected reading of the charge collected by the ion chamber.
    b. C_{TP} corrects for temperature and pressure variations at the time of calibration (see *Temperature and Pressure* on page 49).
    c. P_{elec} corrects for the inaccuracy of the electrometer if calibrated separately.
    d. P_{pol} corrects for chamber polarity effects (see *Polarity Effects* on page 50).

2. N_{gas} represents a calibration of the cavity in terms of absorbed dose to the chamber gas per unit charge or electrometer reading.
    a. N_{gas} is determined using ^{60}Co and normally is provided by a calibration laboratory.

b. N_{gas} is specific to an ionization chamber, independent of the environment surrounding the chamber, and applicable to ionizing radiations where W/e for air is constant (which is true for x-rays and electrons with energies under consideration).

3. $(L/\rho)_{air}^{med}$ is the mean ratio of the restricted mass stopping powers (see *Spencer-Attix Formulation of Bragg-Gray Relationship* on page 53).

4. P_{ion} corrects for ionization recombination losses that occur at the time of calibration and is the inverse of the ionization collection efficiency (see *Recombination* on page 48).

5. P_{repl} is a replacement factor that corrects for perturbation resulting from the chamber being inserted into the medium.
   a. P_{repl} includes a gradient correction to account for the placement of the chamber on the descending portion of the depth dose curve (as recommended) and the cylindrical nature and size of the chamber.
   b. P_{repl} does not include an electron fluence correction if the chamber is placed deeper than the depth of maximum dose [see *Depth of Maximum Dose (d_{max})* on page 61] along the central axis and in a region of transient electronic equilibrium [see *Electronic Equilibrium (EE)* on page 49].

6. P_{wall} accounts for the attenuation and scatter caused by the wall of the chamber being a different material than the medium.

7. $(\mu_{ab}/\rho)_{med}^{water}$ is the scaling factor for a phantom material other than water when the spectral distribution and fluence of primary and scattered photons are equivalent (can require an adjustment of the source-to-detector distance (SDD) and chamber depth in the phantom).

8. SC corrects for the different amount of photon scatter in phantom materials such as polystyrene and acrylic than in water.

9. Recommended depth of calibration is beyond the buildup region and on the exponentially decreasing portion of the depth-dose curve.
   a. A depth of 5 cm is recommended for photon energies <15 MeV.
   b. A depth of 7 cm is recommended for photon energies between 16 MeV and 25 MeV.
   c. A depth of 10 cm is recommended for photon energies from 26 MeV to 50 MeV.

Absorbed Dose from Electrons

$$D_{water} = M \cdot N_{gas} \cdot (L/\rho)^{med}_{air} \cdot P_{ion} \cdot P_{repl} \cdot P_{wall} \cdot (S/\rho)^{water}_{med} \cdot \phi^{water}_{med}$$

1. P_{repl} is a replacement factor for the chamber being inserted into the medium.
   a. P_{repl} does not include a gradient correction if the effective point of measurement of a chamber is placed at the d_{max} (as recommended) for the electron beam.
   b. P_{repl} does include an electron fluence correction.
   c. $P_{repl} = 1$ for plane-parallel chambers because in-scatter, obliquity, and effective point of measurements effects are not applicable or ignored.

2. P_{wall} is assumed to be 1 if the thickness of the chamber wall is small (<0.5 mm) because theoretically there is little perturbation of the electron fluence introduced at this size.

3. $(S/\rho)^{water}_{med}$ is the calibration scaling factor for a phantom material other than water when the spectral distributions are the same at the chamber depth in the phantom (d_{max} recommended).

4. ϕ^{water}_{med} is the scaling correction factor for a phantom material other than water if the electron fluences at the depth of measurement are different.

5. The recommended depth of calibration is d_{max} for any electron energy and for both plastic and water phantoms.

Absorbed Dose in a Medium (TG-51 Calibration Protocol)

Absorbed Dose from Photons

$$D_{Q,water} = M \cdot k_Q \cdot {}_{D,w}N^{Co}$$

1. $D_{Q,water}$ is the absorbed dose to water at the point of measurement of the ion chamber placed under reference conditions for a clinical beam with beam quality Q.

   a. The beam quality Q is specified by the $FDD(10)_X$, which is the photon (X) component of the FDD [see *Fractional Depth Dose (FDD)* on page 64] at a 10 cm depth for a field size of 10 cm × 10 cm on the surface of a phantom at an SSD of 100 cm.

   b. For photon energies <10 MeV, $FDD(10)_X$ is taken as fractional depth dose taken at 10 cm depth or FDD(10).

   c. For photon energies >10 MeV, $FDD(10)_X$ is obtained by taking depth-dose measurements using a 1 mm lead foil placed 30 or 50 cm from the phantom surface and applying equations in the protocol or by using a general formula to correct for electron contamination.

2. $M = M_{raw} \cdot P_{ion} \cdot C_{TP} \cdot P_{elec} \cdot P_{pol}$, which is the charge collected in a measuring chamber in coulombs corrected for recombination effects, temperature and pressure variations, electrometer inaccuracies, and polarity effects.

   a. M_{raw} is the uncorrected reading of the charge collected by the ion chamber.

   b. P_{ion} corrects for ionization recombination losses that occur at the time of calibration (see *Recombination* on page 48).

   c. C_{TP} corrects for temperature and pressure variations at the time of calibration (see *Temperature and Pressure* on page 49).

   d. P_{elec} corrects for the inaccuracy of the electrometer if calibrated separately.

   e. P_{pol} corrects for chamber polarity effects (see *Polarity Effects* on page 50).

3. k_Q is the quality conversion factor which converts the calibration factor for a ^{60}Co beam to that for a photon beam of quality Q.

   a. Values of k_Q as a function of Q are given in the protocol for many ion chambers.

Chapter 7 - Measurement of Ionizing Radiation and Absorbed Dose

  b. Values of k_Q for plane-parallel chambers have not been determined because there is insufficient information about wall correction factors in photon beams other than in ^{60}Co beams.

4. $_{D,w}N^{Co}$ is the absorbed-dose-to-water calibration factor traceable to national primary standards for the ion chamber used.

5. Clinical reference dosimetry geometry:
   a. All dosimetry measurements are performed in a water phantom.
   b. The reference depth for calibration purposes is at 10 cm.
   c. The reference dosimetry is performed in either an SSD or SAD setup, with a 10 cm × 10 cm field size defined on the surface for an SSD setup and defined at the depth of the detector for an SAD setup.

Absorbed Dose from Electrons

$$D_{Q,water} = M \cdot k_{Qe} \cdot {_{D,w}N^{Co}}$$

1. $D_{Q,water}$ is the absorbed dose to water at the point of measurement of the ion chamber placed under reference conditions for a clinical beam with beam quality Q.

   a. The beam quality Q is specified by R_{50}, the depth at which the absorbed dose falls to 50% of the maximum dose in a beam with a field size ≥10 cm × 10 cm on the surface of the phantom for R_{50} <8.5 cm, and ≥20 cm × 20 cm for R_{50} >8.5 cm at an SSD of 100 cm.

   b. R_{50} is determined directly from the measured value of I_{50}, the depth at which the ionization falls to 50% of its maximum value, using the equations:

   $R_{50} = 1.029 \cdot I_{50} - 0.06$ cm (for 2 cm ≤ I_{50} ≤ 10 cm) or

   $R_{50} = 1.059 \cdot I_{50} - 0.37$ cm (for I_{50} >10 cm).

2. $M = M_{raw} \cdot P_{ion} \cdot C_{TP} \cdot P_{elec} \cdot P_{pol}$ which is the corrected charge collected in a measuring chamber in coulombs (see *Absorbed Dose from Photons* on page 56).

3. k_{Qe} is the quality conversion factor that converts the calibration factor for a ^{60}Co beam to that for an electron beam of any beam quality Q and is determined by: $k_{Qe} = P^Q \cdot k'_{R50} \cdot k_{ecal}$.

   a. P^Q corrects for gradient effects and is required for cylindrical chambers only.

   b. P^Q is determined by the ratio of uncorrected ionization currents measured at two depths: $P^Q = M_{raw}(d_{ref} + 0.5 \cdot r_{cav}) / M_{raw}(d_{ref})$

      where

      d_{ref} is reference depth in cm defined below and

      r_{cav} is the radius of the chamber's cavity in cm.

   c. k_{ecal} is the photon-electron conversion factor that converts $_{D,w}N^{Co}$ into $_{D,w}N^{Qecal}$, an electron beam absorbed-dose calibration factor for a specific beam quality Q_{ecal}.

   d. k_{ecal} is fixed for a given chamber model. Values for many ion chambers for a Q_{ecal} of $R_{50} = 7.5$ cm are given in the protocol (k_{ecal} is difficult to measure or calculate for plane-parallel chambers because there is evidence that minor construction details cause relatively large changes in this factor).

   e. k'_{R50} is the electron quality conversion factor that converts $_{D,w}N^{Qecal}$ into $_{D,w}N^Q$, an electron beam absorbed-dose calibration factor for the particular beam quality Q.

   f. k'_{R50} is a function of the electron beam quality specified by R_{50}. Values may be obtained from figures for many ion chambers given in the protocol.

4. $_{D,w}N^{Co}$ is the absorbed-dose-to-water calibration factor traceable to national primary standards for the ion chamber used.

5. Clinical reference dosimetry geometry:

   a. All dosimetry measurements are performed in a water phantom.

   b. The reference depth for calibration purposes is at $d_{ref} = 0.6 \cdot R_{50} - 0.1$ cm.

   c. The reference dosimetry is performed with a field size ≥ 10 cm $\times$ 10 cm for $R_{50} < 8.5$ cm and ≥ 20 cm $\times$ 20 cm for $R_{50} > 8.5$ cm defined on the surface with an SSD setup between 90 and 110 cm.

Other Means To Measure Absorbed Dose

Absolute Measurement of Dose

1. Calorimetry: Measure of energy absorbed in a medium from radiation by detecting the energy transferred to the medium in the form of heat.
   a. Only direct method of measuring absorbed dose.
   b. The rise in temperature of an isolated mass of the medium is measured with a Wheatstone bridge.

2. Ferrous-sulfate (Fricke) Chemical Dosimeter
   a. Most common type of chemical dosimeter.
   b. Ferrous changes to ferric ions when exposed to ionizing radiation (G value is the yield of ferric ions).
   c. An ultraviolet spectrometer is used to measure the change in optical density of the solution, which is linear from about 4 Gy up to 50 Gy.

Relative Measurement of Dose

1. Thermoluminescence Dosimetry (TLD): Electrons from valence band "jump" to conduction band by absorbing energy from ionizing radiation impinging on thermoluminescence crystals. As the electrons fall back to the valence band, some of them get "trapped" in intermediate levels produced by impurities in the crystals. The electrons remain trapped in the intermediate levels until heat energy is applied resulting in the emission of light when the electrons drop to the valence band.
   a. The quantity of light emitted is proportional to the energy absorbed, and thus to dose.
   b. Lithium fluoride is the most commonly used material for thermoluminescence dosimeters.

2. Film Dosimetry: The silver in film consisting of cellulose acetate coated with silver bromide emulsion is reduced to small crystals that change the optical density of the film.
   a. Film provides a very useful method for relative dosimetry with high spatial resolution.

b. $OD = \log_{10}(I_0 / I_t)$ or $I/I_0 = 10^{-OD}$

where

I_0 is the incident radiation, and

I_t is the transmitted radiation.

Typical Dose Calibration Uncertainties

1. Ionization chamber calibration lab: 1% – 2%
2. Fricke dosimeter: 3% – 5%
3. Solid State Methods: 3% – 5%
4. Film: 5% – 10%

CHAPTER 8

RADIATION THERAPY TREATMENT PLANNING

Dose Distribution

Depth of Maximum Dose (d_{max})

Depth of Maximum Dose (d_{max}) is the depth of peak absorbed dose along the central axis.

1. ICRU (International Commission for Radiation Units and Measurements) recommends that the term peak dose (d_{peak}) along the central ray for a single field be used in place of d_{max}, which the ICRU uses to define the point of highest dose, on or off the central axis, for the beam.
2. Depth of maximum dose generally decreases with field size.
3. Approximate depths of maximum dose for various photon energies:

Energy (MeV)	d_{max} (cm)
^{137}Cs (0.662)	0.1
^{60}Co (1.25)	0.5
4	1.0
6	1.5
10	2.0
18	3.3
24	4.0

Photon Equivalent Squares

1. Rectangular approximation (4 A / P): equivalent square ≈ 4 × Area / Perimeter.
2. Accuracy of 4 A / P increases with energy and decreases with depth and with in-field blocking.

Geometric Penumbra

Geometric Penumbra is the gradient of dose across a field edge created by the (effective) size of the source and the beam collimating system.

1. At a depth of d: $P_{geometric} = SD \cdot (SSD + d - SCD) / SCD$

    where

    SD is the source diameter,

    SSD is the source-to-surface distance, and

    SCD is the source-to-collimator distance.

2. If collimation is located at 1/2 SSD, then penumbra at SSD is ≈ source diameter.

3. Penumbra increases as depth increases, SSD increases, and source size increases.

Flatness and Symmetry

1. Flatness: The variation of dose relative to the central axis dose measured over 80% of the field size (defined at 10-cm depth for photons and d_{peak} or just beyond for electrons).

2. Uniformity index: Defined (by the ICRU) for electron beams as the ratio of the area where the dose exceeds 90% of the dose in a reference plane and at a reference depth at the central axis to the geometric cross-sectional area at the phantom surface (should exceed 0.8 at d_{peak} for a 10 cm × 10 cm field).

3. Symmetry: A comparison of dose on one side of the central axis to dose on the opposite side.

Electron Dose Distribution

1. Benefits from a rapid fall-off and finite range that can protect structures below a target and deliver a relatively uniform dose superficially.

2. Surface Dose

    a. Surface dose increases with increasing energy.

    b. For field sizes greater than the practical range (see *Range* on page 32) of the electron beam, the surface dose increases with increasing cone and field size.

    c. For field sizes smaller than the practical range, the surface dose increases with decreasing cone and field size.

Photon Beam Correction Factors

Scatter Factors

1. Collimator Scatter Factor (S_c): At a reference distance from the source, the ratio of the output measured in air along the central axis for a given field size to that for a reference field and depth.

2. Phantom Scatter Factor (S_p): At a reference distance from the source, the ratio of the dose due to photon scatter in a phantom for a given field size to that for a reference field.

3. Total Scatter Correction Factor (S_{cp}) = $S_c \cdot S_p$.

4. Sample Scatter Factors for ^{60}Co normalized to a 10 cm × 10 cm field:

Field Size (cm)	S_{cp}
5 × 5	0.96
30 × 30	1.07

Transmission Factors

1. Tray Factor (TF): Factor that accounts for the reduction in photon radiation output caused by the placement of a plastic tray in the collimator system usually used to hold blocks in place.

2. Wedge Factor (WF): Factor that accounts for the reduction in photon radiation output caused by the placement of a physical wedge filter between the beam source and the patient.
   a. WF varies with wedge angle.
   b. WF varies with depth of measurement and field size.

Dosimetric Calculations

Fractional Depth Dose (FDD)

$$FDD = \left[\frac{\text{Dose at depth in medium}}{\text{Dose at } d_{max} \text{ in medium}}\right] = \left[\frac{D_d}{D_{max}}\right]$$

1. Measured with field size defined on phantom surface.

2. Useful for calculations of SSD setups.

3. Mayneord "F" factor (F_M) applied to correct for treatments at SSD other than standard SSD: FDD @ SSD_2 = F_M x FDD_1 @ SSD_1
   where

$$F_M = \left[\frac{(SSD_1 + d)}{(SSD_1 + d_{max})}\right]^2 \times \left[\frac{(SSD_2 + d_{max})}{(SSD_2 + d)}\right]^2.$$

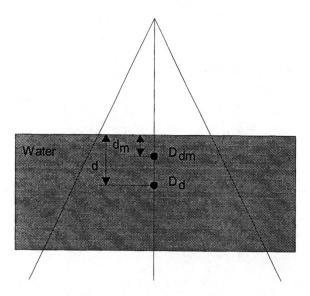

Figure 8.1. Fractional Depth Dose (FDD)

4. Behavior:
   a. FDD decreases with increase in depth after peak (due to exponential attenuation and inverse square law).
   b. FDD increases with increase in photon energy beyond d_{max} (FDD decreases with increasing photon energy between the surface and d_{max}).

c. FDD increases with increase in field size (rapidly at first and then more slowly due to dependence on scatter).
  d. FDD increases with increase in SSD (due to inverse square law).

Tissue-Air Ratio (TAR)

$$TAR = \left[\frac{Dose\ at\ depth\ in\ medium}{Dose\ at\ same\ point\ in\ free\ space}\right] = \left[\frac{D_d}{D_{fs}}\right]$$

1. The dose in free space is measured at a fixed source-to-chamber distance with respect to the dose in medium but with build-up just sufficient to produce electronic equilibrium [see *Electronic Equilibrium* (*EE*) on page 49].

2. Useful for calculations of constant source-to-axis distance (SAD) setups such as with isocentric or rotational therapy calculations.

3. Behavior:
    a. TAR decreases with increase in depth beyond d_{peak}.
    b. TAR increases with increase in energy beyond d_{peak}.
    c. TAR increases with increase in field size beyond d_{peak}.
    d. TAR is independent of SSD for lower-energy megavoltage beams, but becomes dependent on SSD for higher-energy megavoltage beams because of electron contamination.

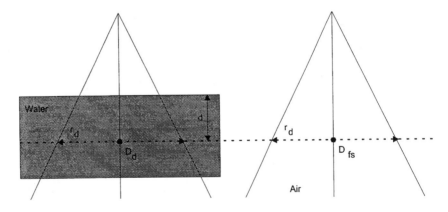

Figure 8.2. Tissue-Air Ratio

Peakscatter Factor (PSF)

$$PSF = \left[\frac{\text{Dose at } d_{max} \text{ in medium}}{\text{Dose at same point in free space}}\right] = \left[\frac{D_{max}}{D_{fs}}\right]$$

1. Similar behavior as TAR except PSF decreases with an increase in energy beyond d_{peak}.

2. Sample PSFs for various photon energies with a 10 cm × 10 cm field:

Energy (MeV)	PSF
0.1	1.2
^{60}Co (1.25)	1.04
10	1.02
20	1.01

Scatter-Air Ratio (SAR)

$SAR = SAR\ (d, r_d) = TAR\ (d, r_d) - TAR\ (d, 0)$

where

r_d is the field size at a depth d.

1. Used to calculate scattered dose in the medium.

2. Can give scatter information under a block:
   Scatter (under block) = SAR (open field) – SAR (blocked field).

3. Clarkson's Method uses TAR and SAR concepts to calculate the dose in an irregular field on the principle that the scatter component of the depth dose (dependent on the field size and shape) can be calculated separately from the primary component (independent of the field size and shape).
   a. The TAR at a point for an irregular field is determined by summing the TAR for the 0 cm × 0 cm field size (primary component) and the averaged SAR for the distances from the calculation point to the field edge measured at ~10 degree increments (scatter component).

b. Requires the extrapolation of beam data to smaller field sizes than measured.

c. Not accurate in penumbral regions.

4. Behavior:

   a. Fraction of dose due to scatter will increase with depth and field size.

   b. SAR increases as radius from point of calculation increases.

Tissue-Maximum Ratio (TMR)

$$TMR = \left[\frac{Dose\ at\ depth\ in\ medium}{Dose\ at\ d_{max}\ in\ medium}\right] = \left[\frac{D_d}{D_{max}}\right]$$

1. Measured with fixed source-to-chamber distance.

2. Useful for calculations of SAD setups.

3. A rough approximation of the dose under a block can be calculated using TMRs:

   TMR (under block) = [block transmission · TMR (blocked field)] +

   [TMR (open field) − TMR (blocked field)]

   where

   TMR (blocked field) represents the TMR for the area under the block.

4. Tissue-Phantom Ratio (TPR) is TMR when a reference depth (d_{ref}) is used in place of d_{max}.

$$TPR = \left[\frac{Dose\ at\ depth\ in\ medium}{Dose\ at\ d_{ref}\ in\ medium}\right] = \left[\frac{D_d}{D_{ref}}\right]$$

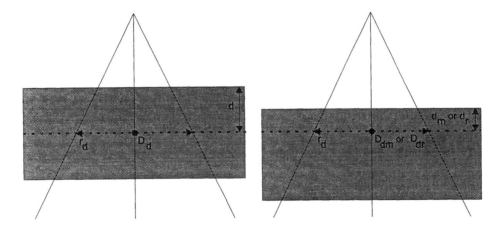

Figure 8.3. Tissue Maximum Ratio (TMR) / Tissue Phantom Ratio (TPR)

5. Both TMR and TPR are based on the assumption that the fractional scatter contribution to the depth dose at a point is independent of the divergence of the beam and depends only on the field size at the point and the depth of the overlying tissue.

6. Behavior:
   a. TMR and TPR decrease with increase in depth beyond d_{peak}.
   b. TMR and TPR increase with increase in energy beyond d_{peak}.
   c. TMR and TPR increase with increase in field size beyond d_{peak}.

Linear Accelerator Monitor Unit Calculations

SSD Calculations

1. At calibration SSD:

   $$\text{Monitor Unit (MU)} = \frac{Dose}{\left(FDD \times D_{ref} \times S_c \times S_p \times TF \times WF\right)}$$

   a. D_{ref} is the dose rate (cGy / MU) at a reference depth, field size, and source-to-chamber distance (reference dose rate).
   b. S_c applies to the field sizes defined at the SAD.
   c. S_p applies to the field size at the SSD.

2. At any SSD:

   FDD must be modified by applying F_M and D_{ref} must be corrected for the change in output at the new SSD:

   $$D_{ref} @ SSD_2 = \left[\frac{D_{ref} @ SSD_1}{PSF(r_{ref})}\right] \times \left[\frac{(SSD_1 + d_{max})}{(SSD_2 + d_{max})}\right]^2 \times PSF(r'_{ref})$$

   where

   r_{ref} is the reference field size and

   $r'_{ref} = r_{ref} \times (SSD_2 / SSD_1)$.

SAD Calculations

1. At calibration SAD:

    a. $\text{MU} = \dfrac{Dose}{\left(TAR \times D_{ref} \times S_c \times TF \times WF\right)}$

    where

    D_{ref} is the reference dose rate measured in air at the SAD, and

    S_c applies to the field sizes defined at the SAD.

    b. $\text{MU} = \dfrac{Dose}{\left[TMR\left(or\ TPR\right) \times D_{ref} \times S_c \times S_p \times TF \times WF\right]}$

    where

    D_{ref} is the reference dose rate measured in a phantom under the reference conditions and

    S_p applies to the field size at the SSD.

2. At any SAD:

    D_{ref} must be corrected for the change in output at the new SAD using the equation above for TMR (or TPR), or using:

    $$D_{ref}\ @\ SAD_2 = D_{ref}\ @\ SAD_1 \times \left(\dfrac{SAD_1}{SAD_2}\right)^2.$$

Photon Dose Calculation Algorithms

Dose Computation Based on Correcting Measured Dose

Relative isodose distributions can be computed by combining tabulated central axis depth doses and dose profile measurements.

1. Measured dose distributions can be corrected for beam obliquity; specific surface contours; beam modifiers such as blocks, wedges, and tissue compensators; and tissue heterogeneities.

2. Correction methods are often based on the separation of primary and scatter contributions based on the assumption that the primary component of dose does not change with field size under electronic equilibrium conditions [see *Electronic Equilibrium (EE)* on page 49].

Model-Based Dose Computation

1. Convolution Superposition: Dose is calculated by modeling the incident energy fluence as it exits the accelerator head and projecting it through a patient to compute the total energy released per unit mass (TERMA), which is then three-dimensionally superposed with an energy deposition kernel.
   a. The incident energy fluence is adjusted to account for the flattening filter, the accelerator head, and beam modifiers.
   b. The TERMA volume is computed by projecting the adjusted incident energy fluence through the patient density volume using a ray-tracing technique.
   c. The energy deposition kernels for monoenergetic photons are generated by Monte Carlo techniques and combined to represent the spread of energy from a primary photon interaction site throughout a volume.

2. Monte Carlo: Dose is calculated by generating a case history of photon interactions that begins with randomly selecting the photon penetration through a medium (from zero to infinity), randomly selecting the energy transfer or dose deposition at the interaction site (from zero to all the photon energy), and then randomly selecting the direction the scattered photon with any remaining energy will travel.
   a. Monte Carlo simulation is capable of computing the dose near interfaces of very dissimilar atomic number materials with very high accuracy.
   b. The accuracy of the calculations is only dependent upon the number of collision histories and the completeness of the simulation model. Computation time increases with the number of histories.

Pencil-Beam Electron Dose Calculation Algorithm

Electron dose distributions for many different setup situations may be calculated using an electron pencil-beam algorithm, which models a broad electron beam as many small individual pencil beams and then adjusting the properties of the pencil beam until the distributions most closely represent measured dose distributions.

1. A pencil beam's incident energy is typically represented by the most probable incident electron energy as determined in *Most Probable Incident Energy* on page 34.
2. The angular scattering of the electrons in air is represented by the parameter sigma-theta-x ($\sigma_{\theta x}$) and can be calculated from measured radiation penumbras.
3. A root mean square scatter correction factor is applied to account for the scattering of the electrons and the spreading out of dose in the computation medium relative to air.
4. Other factors, such as off-axis ratios, may be used to shape the incident fluence of the electron beam and compensate for beam asymmetries.

Field Matching for Photon Beams

Separation of Adjacent Fields

$$\text{Skin Gap} = 1/2 \cdot d \cdot \left(\frac{L_1}{SSD_1} + \frac{L_2}{SSD_2} \right)$$

where

L is the respective field length and

d is the depth at which the fields are to be matched.

1. An extended SSD of one or both fields does not affect the gap because the L/SSD ratio stays the same.
2. A SAD can be used for isocentric setups as long as L corresponds to the field length at SAD.
3. This technique can be a potentially dangerous way to set up fields because of extremely high dose regions just below the match point and extremely low doses just above it.

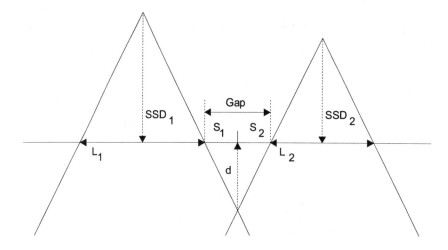

Figure 8.4. Gap Calculation Geometry

Craniospinal Fields

The couch and collimator can be rotated to match diverging spinal fields with diverging cranial fields.

1. Couch rotation: $\theta_{couch} = \tan^{-1}(0.5 \cdot L_c \cdot \sec(\theta_{coll}) / SSD)$

    where

    L_c is the length of the lateral cranial field.

2. Collimator rotation: $\theta_{coll} = \tan^{-1}(0.5 \cdot L_s / SSD)$

    where

    L_s is the length of the posterior spinal field.

Matching of Fields without Divergence

Requires that the fields be angled such that the sides of the matched field are coincident, the use of half-beam blocks facilitates in making edges non-divergent and coincident.

Wedged Fields

A helpful guideline for field matching with wedged pairs comes from an optimum relationship between the wedge angle (WA) and the hinge angle (HA) that will provide a nearly uniform distribution of radiation dose: $WA = 90° - (HA / 2)$.

Heterogeneity Corrections

One-dimensional Heterogeneity Corrections for Photon Beams

Methods that empirically correct for scatter but fail to account for position of the inhomogeneity and its lateral extent.

1. Effective Attenuation Method:

    a. Simple: $D_{Hetero} = D_{Homo} \cdot \mu \cdot (d - d')$

    where

    > D_{Hetero} is the dose to a calculation point beyond an inhomogeneity,
    >
    > D_{Homo} is the dose the calculation point when the homogeneity is assumed to have a water equivalent density,
    >
    > μ is the effective linear attenuation coefficient (see *Exponential Attenuation* on page 35) of the photon beam,
    >
    > d is the physical depth in the patient, and
    >
    > d' is the adjusted water equivalent depth calculated from the summation of the physical path lengths times the relative electron density of the material they traverse

    b. Improved: $D_{Hetero} = D_{Homo} \cdot e^{[\mu \cdot (d - d')]}$.

2. Effective SSD Method:

$$D_{Hetero} = D_{Homo} \cdot \left[\frac{FDD(d', r_s)}{FDD(d, r_s)} \right] \cdot \left[\frac{(SSD + d')}{(SSD + d)} \right]^2$$

    where

    d' is the corrected water equivalent depth,

    d is the physical depth, and

    r_s is the field size at the SSD.

3. TAR Ratio (RTAR) Methods: Improved over Effective SSD Method by accounting for beam divergence through the use of the field size at depth.

    a. Basic: $D_{Hetero} = D_{Homo} \cdot \left[\dfrac{TAR(d', r_d)}{TAR(d, r_d)} \right]$

b. Simple TAR Batho Power Law: $D_{Hetero} = D_{Homo} \cdot \left[\dfrac{TAR(d'',r_d)}{TAR(d',r_d)}\right]^{\rho e - 1}$

where

    d" is the physical distance from the source side of the inhomogeneity to the point of calculation that must lie beyond the inhomogeneity and in a unit density material,

    d' is the physical distance from the non-source side of the inhomogeneity to the point of calculation,

    r_d is the field size at the physical depth of the point of calculation, and

    ρe is the electron density of the inhomogeneity relative to that of water.

c. Improved TAR Batho Power Law: $D_{Hetero} = D_{Homo} \cdot \left[\dfrac{[TAR(d',r)]^{\rho''-\rho'}}{[TAR(d'',r)]^{1-\rho'}}\right]$

where

    ρ" is the electron density of the material in which the point of calculation lies, and

    ρ' is the electron density of the overlying or subsequent material.

Three-dimensional Heterogeneity Corrections for Photon Beams

1. Equivalent TAR Ratio (ETAR) Method: $D_{Hetero} = D_{Homo} \cdot \dfrac{[TAR(d',r'_{d'})]}{[TAR(d,r_d)]}$

    where

        d' is the water equivalent depth,

        $r'_{d'}$ is the scaled field size calculated from relative electron densities throughout the inhomogeneity,

        d is the actual depth, and

        r is the field size at depth d.

    a. ETAR is actually a quasi-3D method that attempts to account for heterogeneities out of the calculation plane.

    b. The irradiated medium is assumed to be semi-infinite and therefore loss of scatter is often not accounted for, resulting in small overestimates of dose.

2. Fast Fourier Transform (FFT) Convolution Method

    a. The scatter inhomogeneity correction is modeled with a linear approximation such that the scatter dose calculations can be implemented using 3-D FFT convolutions.

b. The first scatter dose kernel calculated for water and used in the FFT convolution is scaled linearly with the density at the inhomogeneity.

3. Scatter Ray Trace Methods
   a. Delta Volume (DV) Method: The dose at a point in a medium is found by determining the calculated primary dose contribution and adding pre-calculated, theoretical first-scatter and multiple-scatter contributions.
   b. Dose Spread Array (DSA) and Differential Pencil Beam (DPB) Methods: The dose to a point is calculated using Monte Carlo generated pencil beam dose spread distributions in water and scaling the primary component of dose according to densities encountered along a ray trace to a primary interaction site, and the scatter components according to the densities encountered from the primary interaction site to the dose point.

4. Convolution/Superposition Method: The Convolution calculation algorithm (see *Model-Based Dose Computation* on page 69) can account for heterogeneities by scaled ray-tracing of the path of primary photons and the scaling of the dose kernel values according to the radiological pathlengths between the primary photon interaction sites and the dose deposition sites.

Heterogeneity Corrections for Electron Beams

1. Coefficient of Equivalent Thickness (CET) Method: A method to calculate dose beyond a large, uniform inhomogeneity by correcting the depth traversed through it by its relative electron density to water and applying the depth dose of the corrected depth.
   a. The effective depth $(d_{eff}) = d - z \cdot (1 - CET)$

      where

      d is the physical depth to the point of calculation,

      z is the thickness of the inhomogeneity, and

      CET is the ratio of the inhomogeneity's electron density to that of water.
   b. The CET Method gives good agreement for bone inhomogeneities because the electron density is not much different than water.

c. CET values for lung are dependent on depth within the lung because of the reduced scatter from the low-density tissue, so empirical equations have been derived to determine CET.

2. Electron Pencil Beam Corrections: The pencil beam algorithm can be adapted to account for heterogeneities by incorporating into the calculations an effective depth, calculated assuming that the linear stopping power of the heterogeneity relative to water is relatively independent of electron energy (see *Pencil-Beam Electron Dose Calculation Algorithm* on page 70).

CHAPTER 9

BRACHYTHERAPY PHYSICS

Applications

Interstitial Implants

Interstitial implants involve the implantation of radioactive seeds or needles directly into the tumor volume or in the immediate vicinity of a lesion.

1. Interstitial implants can be temporary or permanent.
2. Interstitial implants are often used for prostate or gynecologic cancers, and are particularly suited for intraoral and superficial tumors.

Intracavitary Insertions

Intracavitary insertions involve radioactive sources placed in a body cavity in the vicinity of a tumor.

1. All intracavitary insertions are temporary.
2. Sources are placed in an applicator positioned in a body cavity such as the uterus, cervix, vagina, bronchus, or nasal cavity.

Surface Applications

Surface applications involve sources placed slightly above or on the skin.

1. Used to treat non-invasively the region of a superficial tumor or lesions in the eye.
2. Sources are often positioned in a mold that conforms to the treated surface or a plaque placed just above the skin.

Treatment Types

Temporary

An implant that delivers radiation continuously over a period of time and is then removed after a desired dose is given.

1. Pre-loaded Manual Loading: Needles containing radioactive isotopes are surgically placed into tissues and left in place until a desired dose has been delivered.
2. Manual Afterloading: Applicators, such as hollow tubes, needles or catheters (flexible plastic tubes), are inserted in a patient, and then loaded by hand with radioactive sources, and removed when a desired dose has been delivered.
3. Remote Afterloading: Treatment similar to manual afterloading except the source loading is driven by remote control.

Permanent

Radioactive sources are surgically implanted within the treatment volume, permanently remaining in the patient.

Source Strength Specification

Exposure Rate

1. Exposure rate constant (Γ) is the exposure rate in roentgens per hour at a point 1 cm from a 1 mCi point source and has units $[(R \cdot cm^2) / (hr \cdot mCi)]$ or $[(R \cdot m^2) / (hr \cdot Ci)]$.

    a. Γ is defined by the ICRU as: $\Gamma = (d^2 / A) \cdot [dx/dt]$ for photons above a δ cutoff energy where d is the distance from a point source of activity A (see *Activity* on page 8) and exposure rate [dx/dt].

    b. Γ is a function of energy and number of photons emitted per decay of a radioisotope.

2. Exposure rate at a distance: The National Council on Radiation Protection and Measurements (NCRP) recommends specifying the strength of gamma emitters in terms of exposure rate in air at a specified distance such as 1 cm or 1 m.

   a. The exposure rate from a point source can be related to the activity of the source by: Exposure rate = $\Gamma \cdot A / d^2$

   where

   Γ is the exposure rate constant,
   A is the activity of the source, and
   d is the distance from the source.

   b. The exposure rate from a linear source follows the relationship:

   Exposure rate = $(\Gamma \cdot A / L \cdot h) \cdot \int e^{-(\mu' \cdot t / \cos\theta)} d\theta$

   where

   L is the active length of the source,
   μ' is the effective attenuation coefficient for the filter,
   t is the source's wall thickness, and
   h and θ are derived from a linear source geometry.

Apparent Activity

The strength of a radioactive source specified in terms of the exposure rate at a distance of 1 m and determined by the ratio of the exposure rate at 1 m with the exposure rate constant of the unfiltered source at 1 m.

Radium Equivalent

The strength of a radioactive source can be expressed in terms of an effective equivalent mass of radium (or "mg·Ra·eq") where the conversion from activity is made by dividing the exposure rate at 1 m by the exposure rate constant of a point source of a 0.5 mm Pt-filtered radium source at 1 m.

Air Kerma Strength

1. $S_K = K \cdot l^2$

   where

   K is the air kerma rate at a specified distance (usually 1 m).

2. S_K has units [cGy · cm² / h] or [µGy · m² / h].

3. The American Association of Physicists in Medicine (AAPM) and the American Brachytherapy Society recommend using S_K to specify source strength [the National Institute for Standards and Technology (NIST) currently specifies sources in terms of S_K].

Radionuclides

Radium-226

1. ^{226}Ra decays to ^{222}Rn with t_H = 1622 years (see *Half-Life and Average Life* on page 8), and then through other transitions until reaching stable ^{206}Pb.
2. HVL = 12 mm in lead [see *Half-Value Layer (HVL or X_H)* on page 37].
3. Γ = 8.25 R·cm^2/ mg·hr from a 0.5 mm Pt-filtered source (see *Exposure Rate* on page 78).
4. ^{226}Ra decays by alpha and beta emission (0.5 mm Pt filters out these radiations) with accompanying gamma emissions of many energies but with mean energy ≈0.83 MeV (see *Modes of Radioactive Decay* on page 11).

Cesium-137

1. ^{137}Cs decays to ^{137}Ba with t_H = 30 years.
2. HVL = 6 mm in lead.
3. Γ = 3.26 R·cm^2/ mCi · hr for an unfiltered source.
4. ^{137}Cs decays by beta emission and by a gamma emission with energy = 0.662 MeV.
5. ^{137}Cs is a good substitute for radium because it has similar penetration and dose distribution in tissue but is less hazardous and requires less shielding.

Gold-198

1. ^{198}Au decays to ^{198}Hg with t_H = 2.7 days.
2. HVL = 3 mm.
3. Γ = 2.38 R·cm^2/ mCi · hr.
4. ^{198}Au decays by beta emission and internal conversion with conversion electrons at 0.329 and 0.403 MeV and by gamma emission with a useful energy of 0.412 MeV.

5. ^{198}Au is usually used for permanent interstitial implants.

Iridium-192

1. ^{192}Ir decays to ^{192}Pt with t_H = 74.2 days.
2. HVL = 3 mm in lead.
3. Γ = 4.69 R·cm^2/ mCi · hr for an unfiltered source.
4. $\Lambda \approx 1.1$ cGy h^{-1} U^{-1} [value depends on specific source type, see *Task Group 43: Dosimetry Formalism for Interstitial Sources (^{192}Ir, ^{125}I, ^{103}Pd)* on page 84].
5. ^{192}Ir decays by beta emission and by gamma emission with mean energy $\approx$ 0.380 MeV (and by electron capture to form ^{192}Os).
6. ^{192}Ir is used for low dose-rate interstitial implants and high dose-rate remote afterloaders.
7. ^{192}Ir has high specific activity (activity per unit mass).

Iodine-125

1. ^{125}I decays to ^{125}Te with t_H = 60.2 days.
2. HVL = 0.025 mm in lead.
3. Γ = 1.45 R·cm^2/ mCi · hr (1999 TG-43 NIST value).
4. $\Lambda \approx 0.9$ cGy h^{-1} U^{-1} (value depends on specific source type).
5. ^{125}I decays to ^{125}Te by electron capture which spontaneously decays to a ground state with the emission of a 35.5 keV gamma, electron capture and internal conversion processes lead to the emission of characteristic x-rays with energies of 27 to 35 keV.
6. ^{125}I is used for permanent interstitial implants and temporary interstitial and surface applications.

Palladium-103

1. ^{103}Pd decays to ^{103}Rh with t_H = 17 days.
2. HVL = 0.008 mm in lead.
3. Γ = 0.65 R·cm^2/ mCi · hr.
4. $\Lambda \approx 0.75$ cGy h^{-1} U^{-1} (value depends on specific source type).

5. ^{103}Pd decays by electron capture and emits characteristic x-rays with energies of 20 to 23 keV (average energy ≈ 21 keV).
6. ^{103}Pd is used for permanent interstitial implants and temporary interstitial and surface applications.

Strontium-90

1. ^{90}Sr decays to ^{90}Y with t_H = 29.8 years.
2. ^{90}Sr decays by pure beta emission (so no Γ associated with ^{90}Sr) with a maximum beta energy = 0.54 MeV, and is in secular equilibrium with ^{90}Y, which decays by pure beta emission with energy = 2.27 MeV (see *Secular Equilibrium* on page 10).
3. ^{90}Sr is used for superficial applications such as the treatment of pterygium (proliferation of blood vessels in the cornea).

Localization

Orthogonal Films

Establishing the coordinates of the radiation sources and points of interest for dose calculations from images made using two orthogonal x-ray beams.

1. Requires precise knowledge of the imaging geometry including the source-to-image or source-to-isocenter distance and the magnification of the image.
2. Deduction of the coordinates in space requires simultaneous application of the information from both images.

"Stereo" Films

Establishing the coordinates of the radiation sources and points of interest for dose calculations from images made using two parallel x-ray beams displaced linearly.

1. Requires a source-to-film distance greater than 100 cm and a shift greater than 30 cm for the highest accuracy.
2. Identification of the same source on both images is often easier than with orthogonal images, but structures that are not point-like may be more difficult to identify.

Dosimetry

Conventional Point Source Approximation

1. Dose-rate = $A \cdot \Gamma \cdot f_{med} \cdot T(r) \cdot \phi_{an} / d^2$

    where

    f_{med} is the f-factor for the medium (see *f-factor (f)* on page 45),

    $T(r)$ is a function that accounts for the scatter and attenuation of radiation in tissue relative to air, and

    ϕ_{an} is an anisotropy constant.

2. Implies that the dose rate depends only on the distance in tissue from the center of the source (r).

3. To determine dose rate to a point in tissue from a number of discrete radioactive sources, simply sum the dose rates to that point from each individual source.

4. Limited because the approximation is expressed in terms that refer to an idealized point source rather than an actual source, the two-dimensional source anisotropy in sources is not accounted for, and the seed-source geometry and encapsulation are not considered.

Sievert Integral

1. Integrating the exposure rate equation for a linear source (see *Exposure Rate* on page 78) gives the Sievert Integral (SI):

    $$SI(x,y) = \int (\Gamma \cdot A / L \cdot r^2) \cdot e^{-(\mu' \cdot t \cdot \sec\theta)} d\theta = (\Gamma \cdot A / L \cdot h) \cdot \int e^{-(\mu' \cdot t \cdot \sec\theta)} d\theta.$$

    (See Figure 9.1. Note: θ must be redefined by $r = y \cdot \sec\theta$ and $x = y \cdot \tan\theta$, where x is defined by the source axis and y is orthogonal to x.)

2. Incorporated into this equation are inverse square law corrections and filtration corrections for the radiation traveling through the source encapsulation, but not scatter and absorption corrections for radiation passing through the tissue medium. To account for scatter and absorption, and to determine dose, the Sievert integral can be multiplied by f_{med} and $T(r)$.

3. This mode agrees to within 5% when tested against Monte Carlo calculations.

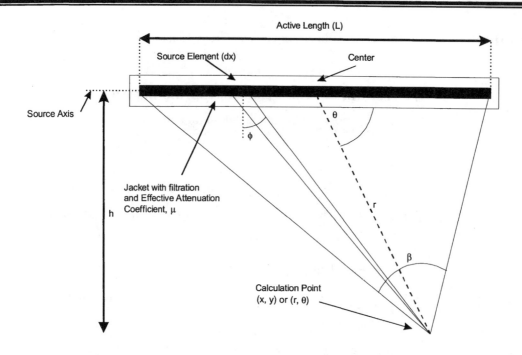

Figure 9.1. Geometrical Relationships for Linear Source Dose Calculation

Task Group 43: Dosimetry Formalism for Interstitial Sources (^{192}Ir, ^{125}I, ^{103}Pd)

Doserate $(r,\theta) = S_k \cdot \Lambda \cdot [G(r,\theta)/G(1,\pi/2)] \cdot F(r,\theta) \cdot g(r)$

where

r and θ are defined in Figure 9.1.

1. S_k is the air kerma strength of the source in µGy m² h⁻¹, which is equivalent to cGy cm² h⁻¹ (1 unit of air kerma strength = 1 U = 1 cGy cm² h⁻¹).

2. Λ is the dose rate constant (the dose rate at 1 cm per U) for the source in tissue medium.

3. $G(r, \theta)$ is a radioactive source geometrical factor that describes the effects on dose caused by the geometry of the source and the distance to the calculation point;

   for a point source $G(r,\theta) = 1/r^2$;

   for a line source $G(r,\theta) = \beta / (L \cdot r \cdot \sin\theta)$.

4. F(r, θ) is a dimensionless anisotropy function that accounts for photon attenuation and scatter in the source encapsulation and medium in the directions other than along the perpendicular bisector of the source axis.
5. g(r) is a dimensionless radial dose function that accounts for photon attenuation and scatter in the medium in the direction of the perpendicular bisector of the source axis.

Implant Dosing Systems

Interstitial Systems

1. Paterson-Parker, Manchester
   a. Characterized by a non-uniform distribution of sources resulting in a uniform distribution of dose.
   b. Source arrangement follows rules to deliver a radiation dose that varies by less than ±10% (specified dose is approximately 10% higher than the dose at the periphery).

2. Quimby
   a. Characterized by a uniform distribution of uniform-strength sources resulting in a non-uniform distribution of dose.
   b. Dose is specified 3 mm beyond the periphery for volumes, and in the center (at the maximum) for planes.
   c. Delivers a higher dose to the middle of a treatment volume than along the edges.

3. Memorial
   a. Based on sources of uniform strength and distribution (similar to Quimby) and characterized by specifying the dose to the implant by the lowest dose found on the edges (known as the minimum peripheral dose).
   b. Uses a "nomograph" (or "nomogram") to determine amount of activity to be implanted based on the premise that the radiation tolerance of tissues in a treatment volume depends on the size of the implant.

4. Paris
   a. Characterized by uniformly spaced source lines of equal length and strength, assigning wider spacing for longer sources or larger treatment volumes, and implanting in one or more parallel and uniformly spaced planes.
   b. Dose prescriptions (reference doses) are to 85% of the average of the local dose minima between neighboring needles (basal dose), providing good dose uniformity within the treatment volume but including a significant volume of normal tissue.
   c. The dose rate is only alterable by varying the source activity along an entire source line.

Intercavitary Systems (Cervical Cancer)

1. Manchester System
   a. Dose is specified to Point A, which is thought to be representative of where the uterine artery crosses the ureter, and is located 2 cm superior along the tandem from the superior aspect of the vaginal fornices containing the ovoids and 2 cm perpendicular to the tandem in the lateral direction.
   b. Dose is often also specified to a Point B, assumed to be the location of lymph nodes, and located 2 cm superior along the body axis from the superior aspect of the vaginal fornices containing the ovoids and 5 cm perpendicular to the tandem in the lateral direction.

2. M.D. Anderson System
   a. Standard loadings for the applicators are based on tandem length, ovoid diameter, and the stage and extent of the patient's disease.
   b. Treatment set with the provisions that the total time not exceed a set limit, the dose to the bladder and rectum remain below allowed tolerances, the vaginal surface dose remains below set limits, and the total mg · hr remain below specified values.

CHAPTER 10

RADIATION PROTECTION

Definitions

Equivalent Dose

An average dose multiplied by a radiation weighting factor (RWF), a quantity related to the linear energy transfer [see *Linear Energy Transfer* (*LET*) on page 88] but independent of the irradiated tissue, and having units of the Sievert (Sv) = 100 rem = 1 J/kg.

Radiation Type	RWF
X-rays, γ-rays, electrons	1
Protons and thermal neutrons (<10 keV)	5
Fast neutrons (10 keV – 2 MeV)	10 – 20
High-energy neutrons (2 – 20 MeV)	5 – 10
Heavy charged particles, α-particles	20

Effective Dose

Sum of equivalent doses to organs that have been multiplied by the corresponding tissue weighting factor (TWF), a quantity that is a measure of the radiobiological damage to a tissue from the particular radiation exposure.

Tissue Type	TWF
Gonads	0.2
Red bone marrow, colon, lung, stomach	0.12
Bladder, breast, liver, esophagus, thyroid	0.05
Skin, bone surface	0.01

Committed Equivalent Dose

The Committed Equivalent Dose is the Equivalent Dose integrated over 50 years.

Committed Effective Dose

The Committed Effective Dose is the Effective Dose integrated over 50 years.

Linear Energy Transfer (LET)

Linear Energy Transfer (LET) is the term used to describe the rate at which energy is deposited as a charged particle travels through matter and is expressed in keV / μ.

1. Heavy charged particles have a high LET compared to electrons.
2. LET is a function of the medium and the mass and charge of the charged particle transferring dose to the medium.
3. High LET produces an exponential survival curve and yields survival curves with low values of D_0 (see *Cell Survival Curve* on page 97); low LET produces survival curves with shoulders before the exponential portion and higher values of D_0.
4. The higher the LET, the less likely cellular repair.
5. LETs of various radiations are:

Radiation	LET (keV/μm)
1 MeV Electron	0.25
^{60}Co (1.25 MeV) γ-ray	0.3
3 MeV X-ray	0.3
250 keV X-ray	3
19 MeV Neutron	7
1 keV Electron	12
2.5 MeV Neutron	20
5 MeV α-particle	100

Effective Dose Limits

Areas

1. Radiation area: 0.05 mSv / hr (5 mrem / hr)
2. High radiation area: 1 mSv / hr (100 mrem / hr)

Occupational Limits (NCRP Report No. 116)

Exposures	Effective Dose Equivalent Limits
Whole-body	50 mSv / yr
Infrequent / Planned	100 mSv
Lens of Eye	150 mSv / yr
Organs, Extremities, Skin	500 mSv / yr
Pregnant Worker	5 mSv / term (0.5 mSv / mo)
Lifetime	10 mSv × Age (years)

Public Limits (NCRP Report No. 116)

Exposures	Effective Dose Equivalent Limits
Whole-body	1 mSv / yr
Infrequent / Planned	5 mSv
Extremities, Skin, Lens of Eye	50 mSv / yr

Shielding

Protective barriers must protect against primary radiation (primary barrier) and scatter and leakage radiation (secondary barrier).

Photon Shielding (NCRP Reports No. 49 and No. 51)

1. Primary Barrier: $P = (W \cdot U \cdot T / d^2) \cdot B$

    a. P is the maximum permissible dose equivalent for the area behind the barrier.
    b. W is the workload of the radiation source.

c. U is the use factor or the fraction of time which the radiation source is directed toward the barrier.

Location	Typical Use Factor (U)
Walls	1/4
Floor	1
Ceiling	1/4 - 1/2

d. T is the occupancy factor of the fraction of operating time during which the area to be protected is occupied (not used for calculation of the hourly dose rate).

Type of Occupancy	Typical Occupancy Factors (T)
Full Occupancy (Work areas, offices)	1
Unrestricted Areas	1
Partial Occupancy (Halls, restrooms)	1/4
Occasional Occupancy (Closets, stairways)	1/8 - 1/16

e. d is the distance to the barrier.

f. B is the required transmission factor for the shielding material or the combination of shielding materials (thickness for barrier is determined from graphs of B versus thickness for the specified photon energy).

2. Secondary Scatter Barrier: $P = [\alpha \cdot W \cdot T / (d^2 \cdot d'^2)] \cdot (F/400) \cdot B_S$

   a. α is the fractional scatter at 1m from the scattering medium for beam area of 400 cm^2 incident at the medium ($\approx 0.1\%$).
   b. U is only allowed if the calculation is broken into directional parts.
   c. d is the distance from the radiation source to the scattering medium.
   d. d' is the distance from the scattering medium to the barrier.
   e. F is the area of the beam incident at the scattering medium in centimeters.
   f. B_S is the required transmission factor for the material used as the secondary scatter barrier (thickness for barrier is determined from graphs of B versus thickness that specify the photon energy and scattering angle).

Chapter 10 - Radiation Protection

g. Scatter dominates the secondary barrier calculations at low photon energies.

h. If the calculated leakage barrier thickness (see below) is greater than the calculated scatter barrier thickness by at least a tenth-value layer in thickness, the scatter barrier thickness can often be substituted by the equivalent of one half-value layer thickness added to the leakage barrier thickness.

4. Secondary Leakage Barrier:

$P = (60 \cdot I) \cdot (W \cdot T / d^2) \cdot B_L$ for therapy units <500 kVp

$P = 0.001 \cdot (W \cdot T / d^2) \cdot B_L$ for megavoltage therapy units

a. I is the maximum tube current.

b. B_L is the required transmission factor for the material used as the secondary leakage barrier (thickness for barrier is determined from graphs of B versus thickness that specify the nominal photon energy).

c. Leakage for megavoltage units restricted to ≈ 0.1% at 1 m.

d. d is the distance from the radiation source to the barrier.

e. Leakage dominates the secondary barrier calculations at high photon energies.

f. If the calculated scatter barrier thickness is greater than the calculated. leakage barrier thickness by at least a tenth-value layer in thickness, the leakage barrier thickness can often be substituted by the equivalent of one half-value layer thickness added to the scatter barrier thickness.

Neutron Shielding

1. Required for linear accelerators with photon energy >15 MeV.

2. If concrete is used for the primary and secondary photon barriers, then neutron shielding is mostly a problem only at the entrance to the room.

3. Consists of hydrogenous material to thermalize neutrons (water or plastic), a high-absorption cross-section material to capture thermalized neutrons (cadmium or boron), and a material to stop any gammas that may originate from the capturing process (lead).

4. The fractional scatter for neutrons incident on concrete walls is about 10%.

Monitoring Instruments

Film

1. Good for personnel monitoring.
2. Fairly accurate.
3. Sensitive down to ≈ 200 µGy.
4. Markedly energy dependent.

Solid-State Detectors

1. Thermoluminescence dosimeter (TLD).
   a. Good for personnel monitoring.
   b. Good accuracy.
   c. Have wide sensitive range from 0.1 mGy to 4 Gy.
   d. Energy dependence less than 20%.

2. Scintillation detection.
   a. Good for imaging, spectroscopy, and detecting alpha contamination and low levels of beta and low-energy photon emitters.
   b. Good accuracy.
   c. Very sensitive with limited range.
   d. Inorganic detectors are energy dependent; plastics are almost energy independent.

Gas-Filled Chambers

Based on the principle that as radiation passes through gas, it ionizes atoms in its path causing the liberation of electrons in the chamber that are collected and measured on a positively-charged central electrode.

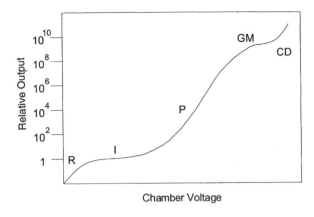

Figure 10.1. Output versus Gas-Filled Chamber Voltage

1. Recombination region (R) occurs when the voltage across the chamber is so low that many electrons and positive ions recombine before collection.

2. Ion chambers (I) are used in the first saturation region where most electrons generated are collected and, therefore, the ionization current is proportional to exposure rate.
   a. Ion chambers measure ionization current.
   b. Response is relative to actual ionization produced.
   c. Good for portable surveys when the exposure rate is greater than 0.1 mR / hr.
   d. Very good accuracy, low sensitivity, wide range, and modest energy dependence.

3. Proportional counters (P) are used in the proportional region where the electrons are generated and then accelerated rapidly toward the central electrode causing further ionization.
   a. Proportional counters count pulses and assess size of pulse.
   b. Response is relative to radiation energy and type of radiation causing the pulse.
   c. Proportional counters are good for assaying radionuclides in a laboratory, accurate with very stable voltage supply to electrodes, sensitive, energy dependent, and have a limited maximum detectable signal.

4. Geiger-Müller counters (G-M) are used in the second saturation region where a single ionizing event will trigger an avalanche of secondary electrons giving a large electron pulses of uniform size.
   a. G-M counters only give counts per unit time, but can be calibrated for a certain energy at a particular part of the response range in mR per hour.
   b. Good for portable surveys and exposures between 0.001 and 10,000 mR/hr, relatively sensitive, very energy dependent.
   c. G-M counters suffer from "dead time" (see *Radioactive Decay Counting* on page 4).
   d. Non-paralyzable dead time refers to the time after an initial electron pulse when an additional ionizing event will not cause an electron avalanche and therefore will have no effect on the counter.
   e. Paralyzable dead time refers to the time after an initial pulse when a sufficiently large electron pulse can be counted, but will result in the subsequent count not being detected but in the generation of a new dead-time period.

5. Continuous discharge region (CD) occurs when the voltage is so high that a single ionizing event will completely discharge the chamber, and a continuous current will be produced as the electrons are stripped from the gas.

CHAPTER 11

RADIOBIOLOGY

Biological Effects of Radiation

Biological Effects

1. Stochastic (with assumption of a linear, no-threshold model).
    a. No associated threshold.
    b. Probability of occurrence increases with dose.
    c. Severity of effect is independent of dose.
    d. Values occur randomly and cannot be predicted.
    e. Example: carcinoma.

2. Non-Stochastic.
    a. Has threshold.
    b. Increases in severity with increase in dose.
    c. Value can be predicted for given conditions.
    d. Examples: cell death, cataracts.

Radiobiological Equivalent (RBE)

$$RBE = \frac{[Dose\ from\ standard\ radiation\ to\ produce\ a\ given\ biological\ effect]}{[Dose\ from\ the\ test\ radiation\ to\ produce\ the\ same\ biological\ effect]}$$

1. 200 keV x-ray radiation from an orthovoltage radiation generator (see *Orthovoltage* on page 23) was originally used to define an RBE of 1.0.
2. Standard radiation is 250 keV x-ray.
3. Used to compare the effects of different types of radiation.
4. Depends on dose, species, and effect used for the determination.

Oxygen Enhancement Ratio (OER)

$$OER = \frac{[\text{Dose to produce a given effect with no } O_2 \text{ present}]}{[\text{Dose to produce the same effect with 1 atm of air}]}$$

1. The presence of oxygen at the time of irradiation acts as a sensitizing agent.
2. Sample OERs for various radiations include:

Radiation Type	OER
X-rays and γ-rays	2 – 3.5
Neutrons	1.5
α-particles	1

Cell Survival

Cell Radiosensitivity

1. Law of Bergonie and Tribondeau: Cells undergoing mitosis, cells with a long dividing time, and undifferentiated cells are the most radiosensitive.
2. The higher the LET [see *Linear Energy Transfer (LET)* on page 88] of the radiation, the less likely cellular repair.

Alpha-Beta Model of Cell Survival: $S = e^{-(\alpha \cdot D + \beta \cdot D^2)}$

1. The ratio α/β is dose at which linear and quadratic terms contribute equally to biological response.
2. The ratio α/β increases with LET, is independent of dose, and is dependent of the biological end point, type of cell, and the cell's environment.
3. Typical α/β for early-responding tissues is ≈10 Gy, and for late-responding tissues ≈3 Gy (tumors tend to behave like early-responding tissue).

Cell Survival Curve

A plot of the surviving fraction of cells versus delivered dose.

1. Extrapolation number (n) represents the number of targets in a cell that must be hit by radiation to cause cell death.
2. D_0 is the mean lethal dose that kills all but 37% of cell population and is an indicator of cell radiosensitivity.
3. D_q is the quasi-threshold dose below which exponential killing is not observed.
4. Single hit survival: $N(D) = N_0 \cdot e^{-D/D_o}$, where N is the initial number of cells.
5. Multi-target, single hit survival: $N(D) = N_0 \cdot [1 - (1 - e^{-D/D_o})^n]$, where n is the number of targets.

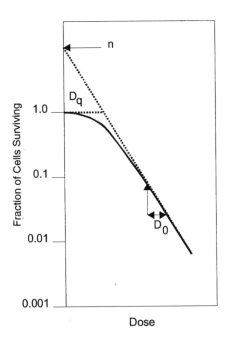

Figure 11.1. Cell Survival Curve

Terms And Acronyms

Term	Definition
A	Mass number
AAPM	American Association of Physicists in Medicine
amu	Atomic mass unit
BE	Binding energy
CD	Continuous discharge
CET	Coefficient of equivalent thickness
CSDA	Continuous slowing down approximation
CT	Computed tomography
DPB	Differential pencil beam
DSA	Dose spread array
DV	Delta volume
E	Relativistic energy/energy
EE	Electronic equilibrium
esu	Electrostatic unit
ETAR	Equivalent TAR
FDD	Fractional depth dose
FFT	Fast Fourier transform
G-M	Geiger-Müller
GR	Grid ratio
HA	Hinge angle
HU	Heat unit
HU	Hounsfield unit
HVL	Half-value layer
ICRU	International Commission on Radiation Protection and Measurements
KE	Kinetic energy
KERMA	Kinetic energy released in the medium

Term	Definition
L/ρ	Restricted mass stopping power
L	Restricted stopping power
LET	Linear energy transfer
LINAC	Linear accelerator
MTF	Modulation transfer function
NCRP	National Council on Radiation Protection and Measurements
NIST	National Institute for Standards and Technology
OER	Oxygen enhancement ratio
PSF	Peakscatter factor
R	Practical range
rad	Radiation absorbed dose
RBE	Radiobiological equivalent
RTAR	TAR ratio
RWF	Radiation weighting factor
S	Stopping power
SAD	Source-to-axis distance
SAR	Scatter-Air ratio
SCD	Source-to-collimator distance
SD	Source diameter
SDD	Source-to-detector distance
SSD	Source-to-surface distance
3-D	Three dimensional
TAR	Tissue-Air ratio
TERMA	Total energy released per unit mass
TF	Tray factor
TG	Task Group
TLD	Thermoluminescence dosimetry
TMR	Tissue-Maximum ratio
TPR	Tissue-Phantom ratio
TWF	Tissue weighting factor
WA	Wedge angle
WF	Wedge factor
Z	Atomic number

Suggested Readings

American Association of Physicists in Medicine (AAPM) Reports:

-- "A protocol for the determination of absorbed dose from high-energy photon and electron beams." Report of AAPM Task Group No. 21. In *Medical Physics* Vol. 10, Issue 6, November/December 1983.

No. 32 "Clinical Electron-Beam Dosimetry." Report of AAPM Radiation Therapy Committee Task Group No. 25. Reprinted from *Medical Physics* Vol. 18, Issue 1, January/February 1991.

No. 46 "Comprehensive QA for Radiation Oncology." Report of AAPM Radiation Therapy Committee Task Group No. 40. Reprinted from *Medical Physics* Vol. 21, Issue 4, April 1994.

No. 47 "AAPM Code of Practice for Radiotherapy Accelerators." Report of AAPM Radiation Therapy Task Group No. 45. Reprinted from *Medical Physics* Vol. 21, Issue 7, July 1994.

No. 51 "Dosimetry of Interstitial Brachytherapy Sources." Recommendations of the AAPM Radiation Therapy Committee Task Group No. 43. Reprinted from *Medical Physics* Vol. 22, Issue 2, February 1995.

No. 59 "Code of Practice for Brachytherapy Physics." Report of AAPM Radiation Therapy Committee Task Group No. 56. Reprinted from *Medical Physics* Vol. 24, Issue 10, October 1997.

No. 61 "High Dose-Rate Brachytherapy Treatment Delivery." Report of AAPM Radiation Therapy Committee Task Group No. 59. Reprinted from *Medical Physics* Vol. 25, Issue 4, April 1998.

No. 67 "Protocol for Clinical Reference Dosimetry of High-Energy Photon and Electron Beams." Report of AAPM Task Group 51. Reprinted from *Medical Physics* Vol. 26, Issue 9, September 1999.

No. 68 "Permanent Prostate Seed Implant Brachytherapy." Report of AAPM Task Group No. 64. Reprinted from *Medical Physics* Vol. 26, Issue 10, October 1999.

Attix, Frank H. *Introduction to Radiological Physics and Dosimetry*. John Wiley & Sons, Inc., New York, NY, 1986.

Bentel, Gunilla C. *Radiation Therapy Planning*, 2nd Ed. McGraw-Hill, Health Professions Divisions, New York, NY, 1996.

Curry, Thomas S., James E. Dowdey, and Robert C. Murry, Jr. *Christensen's Physics of Diagnostic Radiology,* 4th Ed. Lea & Febiger, Philadelphia, PA, 1990.

Hall, Eric J. *Radiobiology for the Radiologist*, 4th Ed. J. B. Lippincott Company, Philadelphia, PA, 1994.

Halliday, David, and Robert Resnic. *Fundamentals of Physics*, 3rd Ed. John Wiley & Sons, Inc., New York, NY, 1988.

Hendee, William R. *Medical Radiation Physics*, 2nd Ed. Yearbook Medical Publishers, Inc., Chicago, IL, 1979.

Hendee, William R. *Radiation Therapy Physics*, 2nd Ed. Yearbook Medical Publishers, Inc., Chicago, IL, 1981.

International Commission on Radiation Units and Measurements (ICRU) Reports:

- No. 33   "Radiation Quantities and Units." ICRU, Washington, DC, 1980. (Out of Print).
- No. 38   "Dose and Volume Specification for Reporting Intracavitary Therapy in Gynecology." ICRU, Washington, DC, 1985.
- No. 50   "Prescribing, Recording, and Reporting Photon Beam Therapy." ICRU, Washington, DC, 1993.

Johns, Harold E., and John R. Cunningham. *The Physics of Radiology*, 4th Ed. Charles C. Thomas, Springfield, IL, 1983.

Karzmark, C. J., Craig S. Nunan, and Eiji Tanabe. *Medical Electron Accelerators*. McGraw-Hill, Inc., Health Professions Division, New York, NY, 1993.

Khan, F. M. *The Physics of Radiation Therapy*. Williams and Wilkins, Baltimore, MD, 1983.

National Council on Radiation Protection and Measurements (NCRP) Reports:

- No. 49   "Structural Shielding Design and Evaluation for Medical Use of X-Rays and Gamma Rays up to 10 MeV." NCRP, Washington, DC, 1976.
- No. 51   "Radiation Protection Design Guidelines for 0.1–100 MeV Particle Accelerator Facilities." NCRP, Washington, DC, 1977.
- No. 91   "Recommendations on Limits for Exposure to Ionizing Radiation." NCRP, Washington, DC, 1987.
- No. 102   "Medical X-Ray, Electron Beam and Gamma Ray Protection for Energies up to 50 MeV." NCRP, Washington, DC, 1989.
- No. 116   "Recommendations of Exposure to Ionizing Radiation." NCRP, Washington, DC, 1993.

Taylor, John R. *An Introduction to Error Analysis*. University Science Books, 1982.